Radiation Heat Transfer

H. R. N. Jones

formerly Senior Lecturer, Department of Chemical and
Process Engineering, University of Sheffield

Series sponsor: AstraZeneca

AstraZeneca is one of the world's leading pharmaceutical companies with a strong
research base. Its skill and innovative ideas in organic chemistry and bioscience create
products designed to fight disease in seven key therapeutic areas: cancer, cardiovascular,
central nervous system, gastrointestinal, infection, pain control, and respiratory.

AstraZeneca was formed through the merger of Astra AB of Sweden and Zeneca Group
PLC of the UK. The company is headquartered in the UK with over 50,000 employees
worldwide. R&D centres of excellence are in Sweden, the UK, and USA with R&D
headquarters in Södertälje, Sweden.

AstraZeneca is committed to the support of education in chemistry and chemical
engineering.

OXFORD
UNIVERSITY PRESS

OXFORD

UNIVERSITY PRESS

Great Clarendon Street, Oxford OX2 6DP

Oxford University Press is a department of the University of Oxford.
It furthers the University's objective of excellence in research, scholarship,
and education by publishing worldwide in

Oxford   New York

Athens   Auckland   Bangkok   Bogotá   Buenos Aires   Calcutta
Cape Town   Chennai   Dar es Salaam   Delhi   Florence   Hong Kong   Istanbul
Karachi   Kuala Lumpur   Madrid   Melbourne   Mexico City   Mumbai
Nairobi   Paris   Saõ Paulo   Singapore   Taipei   Tokyo   Toronto   Warsaw

with associated companies in   Berlin   Ibadan

Oxford is a registered trade mark of Oxford University Press
in the UK and in certain other countries

Published in the United States
by Oxford University Press Inc., New York

A catalogue record for this book is available from the British Library

Library of Congress Cataloging in Publication Data
ISBN 0 19 856455 4

Typeset by EXPO Holdings, Malaysia
Printed in Great Britain
on acid-free paper by
Bath Press, Avon

Series Editor's Foreword

The Oxford Chemistry Primers provide concise introductions to a range of topics encountered by both chemistry and chemical engineering students. The series has been designed to contain only the essential material that would be covered in an 8–10 lecture course. This Primer provides a thorough, yet easy to understand, introduction to heat transfer phenomena by radiation, and complements the earlier Primer of Winterton which considered heat transfer by conduction, forced convection and natural convection with only a brief introduction to the concept of radiation.

Howard Jones has based this text on a popular lecture course he gave for many years in the Department of Chemical Engineering at Cambridge. Throughout the Primer he illustrates the theory with examples that are easy to follow, thereby making this an invaluable text for anyone who wants to be able to solve commonly encountered, relevant problems in radiative heat transfer.

Lynn F. Gladden
Department of Chemical Engineering
University of Cambridge

Author's Preface

When I first started lecturing radiation heat transfer to undergraduate students, it soon became clear that there was not an ideal textbook which could be recommended to them. Textbooks were of two types: there were general engineering heat transfer texts, which often gave a fairly superficial treatment and rarely ventured beyond simple problems of radiation between surfaces, and there were the highly specialised texts, which, while commendably rigorous, went far beyond what an undergraduate student would need. This text attempts to fill the gap by introducing the non-specialist not only to the basics of radiative exchange between surfaces, but also to the complexities of gas radiation and the treatment of combined modes of heat transfer. It also covers a number of practical applications involving radiation heat transfer, including heat transfer in furnaces, techniques for the measurement of temperature, and radiation from combustion gases.

The text works from a student's point of view, and is based firmly in the tradition of hand calculation, as encountered in undergraduate teaching programmes. Modern computational techniques have, of course, transformed heat transfer modelling for the specialist, but it is not the intention to cover those aspects here. The aim is for this text to provide the novice with the necessary grasp of the fundamentals, which can subsequently be extended to more complex geometries.

I should like to thank British Gas plc, whose generous sponsorship of a teaching fellowship at Cambridge University enabled me to develop much of the course material included here. I should also like to thank former colleagues at the University of Sheffield for their support, particularly Dr Peter Foster and Dr Yajue Wu for their comments on the manuscript.

2000   H. R. N. J.

Contents

1 Fundamentals of thermal radiation

1.1 Introduction

Thermal radiation is the third mechanism of heat transfer after conduction and convection. All three mechanisms transfer heat from A to B by virtue of A being at a higher temperature than B. This transfer of heat is a mechanism of energy transfer, the aim of which is to achieve thermal equilibrium and hence equal temperatures. The three mechanisms invoke quite different methods of energy transfer. Both conduction and convection heat transfer between two points rely on the participation of an intervening medium. Conduction from A to B involves the direct exchange of kinetic energy between adjacent particles: a faster moving (i.e. hotter) particle will transfer energy to a slower (colder) particle during a collision. Convection is similar but includes the additional effect of the bulk motion of the particles, e.g. fluid flow near a hot wall, where the particles adjacent to the wall are heated and then carried away in the flow only to be replaced by colder particles.

Radiation heat transfer is very different, in that no intervening medium is required. Thermal radiation is an emission of electromagnetic waves (or photons), and as such can be transferred between two bodies even if they are separated by a vacuum. The emission arises from natural oscillations and energy transitions of the electrons in the molecules. All materials, whether they be gas, liquid, or solid, emit their own radiation and absorb radiation which has been emitted elsewhere. For opaque materials emission of radiation is a surface phenomenon; molecular motion within the material still generates radiation, but this is absorbed by adjacent molecules before it can escape from the surface. For gases and transparent materials, such as glass or water, radiation is a volumetric phenomenon in that the observed emission originates from all parts of the volume of material.

Because thermal radiation is an electromagnetic radiation, we can attribute to it the standard wave properties of velocity (c), frequency (ν) and wavelength (λ), which are related by the equation:

$$c = \nu\lambda \tag{1.1}$$

Transmission through a vacuum (and most gases at normal temperatures and pressures) will be at a velocity of 3.0×10^8 m/s. As the molecular motion giving rise to this emission is thermal in nature, the intensity of the radiation depends on the temperature of the emitting material. We also find that radiation is emitted over a very wide range of wavelengths, and that the radiation is not equally distributed over those wavelengths. The distribution of emitted wavelengths is generally referred to as the *spectral distribution* or *spectrum*.

Another important distinction between radiation and the other two modes of heat transfer is the temperature dependence of the rate of heat transfer. The conductive and convective heat flux (q_{cond} and q_{conv}) between two points of temperature T_1 and T_2 can be expressed by

$$q_{cond} = \frac{k(T_1 - T_2)}{L} \qquad q_{conv} = h(T_1 - T_2)$$

where k is the thermal conductivity of the material, L is the distance between the two points, and h is the convective heat transfer coefficient. Although h and k are often weak functions of temperature, the fluxes are usually considered to have approximately a linear dependence on the temperature difference. Radiative fluxes are quite different in that they are proportional to the difference between the fourth power of the absolute temperatures, i.e.:

$$q_{rad} \propto (T_1^4 - T_2^4)$$

Radiation heat transfer is therefore particularly important at high temperatures, such as those found in combustion systems, solar energy, and nuclear reactors.

1.2   The electromagnetic spectrum

As indicated above, thermal radiation is an electromagnetic emission. The electromagnetic spectrum is generally divided (somewhat arbitrarily) into a number of different regions (Figure 1.1). Note that materials emit radiation across the entire electromagnetic spectrum. However most of the emission is concentrated in the range 10^{-7} to 10^{-4} m, which encompasses the near ultra-violet (UV), all the visible region, and most of the infra-red (IR). Human senses feel incident radiation in this wavelength band as heat. Shorter UV wavelengths are not detected as heat but can do physiological damage (e.g. UV from the sun, X-rays). Within a very narrow range (380 to 760 nm) radiation stimulates the optic nerve, i.e. in this range we can both feel and see thermal radiation emitted from a hot surface.

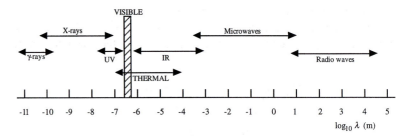

Figure 1.1   Schematic showing the regions of the electromagnetic spectrum.

1.3   Basic definitions

Emissive radiant flux
We define the emissive radiant flux (in W/m^2) as the thermal radiation emitted from a surface per unit time per unit surface area. An alternative, often-used

name is emissive power. This name, however, can be misleading as it disguises the fact that we are dealing with a flux. As indicated above, a wide range of wavelengths is generally emitted from a surface. The intensity of emission at various wavelengths will be discussed in detail later. At this stage it suffices to define two parameters. The *monochromatic emissive flux* (W_λ) is the rate of emission per unit area of radiation in the wavelength interval λ to $\lambda + d\lambda$. We can also define the *total emissive flux* (W) as the total emitted energy summed over all wavelengths. The two are linked by the relation

$$W = \int_0^\infty W_\lambda d\lambda \tag{1.2}$$

If the dependence of W_λ on λ is known, then the integral may be evaluated in order to determine W.

Incident radiant flux

The incident radiant flux (or irradiation) may be defined similarly as the rate at which thermal radiation arrives on a surface per unit surface area. This radiation originates from emissions from other neighbouring surfaces, and therefore contains a wide range of wavelengths, dependent on the temperature of each emitting surface. Again two parameters can be defined. The *monochromatic incident flux* (G_λ) is the incident flux of radiation in the wavelength interval λ to $\lambda + d\lambda$, whereas the *total incident flux* (G) is the total incident energy summed over all wavelengths. The two are linked by the relation

$$G = \int_0^\infty G_\lambda d\lambda \tag{1.3}$$

If the dependence of G_λ on λ is known, then the integral may be evaluated in order to determine the total flux.

Absorptivity, reflectivity and transmissivity

Consider radiation falling on the surface of a body of finite thickness (Figure 1.2). Part of the radiation may be absorbed by the material, part may

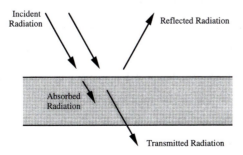

Figure 1.2 Absorption, reflection, and transmission of radiation incident on a body of finite thickness.

be reflected away, and part may be transmitted through the material. The fractions absorbed, reflected, or transmitted are known as the *absorptivity* (α), the *reflectivity* (ρ), and the transmissivity (τ), respectively. The fraction absorbed, reflected, or transmitted by a material is generally dependent on the wavelength of the incident radiation, i.e. there is spectral dependence. Glass offers a good example of spectral dependence in that it is transparent to visible wavelengths, but opaque to infra-red wavelengths greater than around 2.5 μm. A surface may also exhibit directional effects, whereby the surface responds differently to incoming radiation, depending on the angle between the incident beam and the surface. Most engineering calculations assume that surfaces are *diffuse*, i.e. that the absorptivity is independent of direction. (Important exceptions are the reflecting surfaces used in space applications.) Directional dependence is not considered in detail here; the interested reader is referred to the more specialist texts.

We can define *monochromatic* and *total* absorptivity, reflectivity, and transmissivity as follows:

$\alpha_\lambda, \rho_\lambda, \tau_\lambda$ fraction of incident radiation at wavelength λ absorbed/reflected/ transmitted.

α, ρ, τ fraction of incident radiation over all λ absorbed/reflected/ transmitted.

As each of these parameters is a fraction of the incident flux, then each must have a value between 0 and 1. Also, since all incident radiation can only be absorbed, reflected, or transmitted, energy conservation requires that

$$\alpha_\lambda + \rho_\lambda + \tau_\lambda = 1 \quad \text{and} \quad \alpha + \rho + \tau = 1 \tag{1.4}$$

Nearly all engineering solids are opaque, i.e. no radiation is transmitted, in which case

$$\alpha_\lambda + \rho_\lambda = 1 \quad \text{and} \quad \alpha + \rho = 1 \tag{1.5}$$

We will frequently be using this form in heat transfer problems. In fact, this equation is in such general use that it is very easy to forget that it applies only to opaque surfaces.

Radiosity
We define the radiosity (J) of a surface as the total radiative flux leaving the surface. This combines the natural thermal emission from the surface together with any reflected incident radiation. Thus radiosity is given by:

$$J = W + \rho G \tag{1.6}$$

Since W and G have spectral dependence, J must also be wavelength-dependent, so we can define a monochromatic radiosity (J_λ) and a total radiosity (J), which is summed over all wavelengths:

$$J = \int_0^\infty J_\lambda \, d\lambda \tag{1.7}$$

1.4 Black body radiation

When discussing real surfaces it is useful to introduce the concept of the *black body*. A black body is defined as an object or surface which absorbs all incident thermal radiation of all wavelengths, i.e. $\alpha_\lambda = \alpha = 1$; $\rho_\lambda = \rho = 0$; $\tau_\lambda = \tau = 0$. Since no radiation is reflected the only radiation leaving the surface is its own thermal emission, i.e. the radiosity and emissive flux (both monochromatic and total) are equal. In practice perfect black bodies do not exist, although we can get quite close, e.g. a surface can be specially coated with carbon black or platinum black so that it will absorb around 99% of incident radiation. An alternative simulation of a black body is known as the *black body cavity* (Figure 1.3), which consists of a hollow box with a small hole in one side. Any radiation entering the box will impinge on the opposite wall, where a fraction will be reflected and a fraction absorbed. The reflected radiation will then travel to another wall, where part is absorbed, part reflected. This process of internal reflection will continue, with the reflected beam getting progressively weaker and weaker. If the hole through which the radiation entered is sufficiently small then the probability of the radiation escaping is so low that the hole is essentially a total absorber of incoming radiation and appears to behave like a black body.

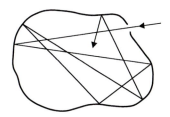

Figure 1.3 Absorption of incident radiation in a black body cavity.

As well as being a perfect absorber of radiation a black surface is also a perfect emitter. By this we mean that for a given temperature no object can emit more radiation (either monochromatic or total) than one which is black. Consider a body emitting and receiving radiation from its surroundings. When the body is a thermal equilibrium with its surroundings the incident flux will equal the radiative flux leaving the surface (i.e. the radiosity). So, using Equation 1.6 we may write:

$$G = J = W + \rho G$$

This equation is valid for any non-black surface of reflectivity ρ. Because the incident flux is constant (at a given temperature) a surface with a low reflectivity will have a high thermal emission W. If the surface in question is black, then $\rho = 0$ and the equation reduces to $G = W_b$, where the subscript b is used to indicate a black body. It follows that the emission from a black body must be the highest possible at a given temperature. This argument is valid both for total and monochromatic radiation.

The notion that a non-black surface emits less than a black surface leads to the concept of emissivity. The monochromatic emissivity (ε_λ) of a body is defined as the ratio of the monochromatic emissive flux at wavelength λ and temperature T to the monochromatic emissive flux from a black body at the same wavelength and temperature.

$$\varepsilon_\lambda = \frac{W_\lambda}{W_{b\lambda}} \tag{1.8}$$

Similarly, we can define the total emissivity as the ratio of the total emissive flux at temperature T to the total emissive flux from a black body at the same temperature:

$$\varepsilon = \frac{W}{W_b} \tag{1.9}$$

Substituting Equation 1.2 followed by Equation 1.8 gives:

$$\varepsilon = \frac{W}{W_b} = \frac{\int\limits_0^\infty W_\lambda \, d\lambda}{\int\limits_0^\infty W_{b\lambda} \, d\lambda} = \frac{\int\limits_0^\infty \varepsilon_\lambda W_{b\lambda} \, d\lambda}{\int\limits_0^\infty W_{b\lambda} \, d\lambda} \tag{1.10}$$

Thus if the spectral variation of black body emission and monochromatic emissivity are known, the total emissivity may be determined. We shall see how this expression can be simplified for practical systems later.

1.5 Spectral distribution of black body radiation

The spectral dependence of black body emission is shown in Figure 1.4, which plots the monochromatic emissive flux ($W_{b\lambda}$) as a function of wavelength for various temperatures. A number of features should be noted:

(i) $W_{b\lambda}$ shows a clear dependence both on wavelength and on the temperature of the emitter.

(ii) For any given wavelength λ, $W_{b\lambda}$ increases with temperature, i.e. isotherms do not cross.

(iii) For any given temperature, $W_{b\lambda}$ passes through a maximum, which moves to a lower wavelength as temperature increases.

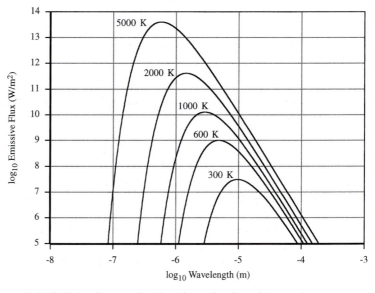

Figure 1.4 Emissive flux as a function of wavelength and temperature.

(iv) From Equation 1.2, the total emission W_b at a particular temperature is the area under the curve for that temperature, i.e.:

$$W_b = \int_0^\infty W_{b\lambda}\,d\lambda \qquad (1.11)$$

A theoretical explanation of black body emission was sought by many physicists at the end of the nineteenth century. Using classical principles of statistical thermodynamics and the kinetic theory of gases, Rayleigh and Jeans derived an expression for $W_{b\lambda}$:

$$W_{b\lambda} = \frac{2\pi ckT}{\lambda^4} \qquad (1.12)$$

where k is the Boltzmann constant. This produces fairly good agreement with experiment at longer wavelengths, and, indeed, is still in use in radio engineering. However, it is not recommended for use in most engineering applications; its mathematical form clearly does not predict a maximum, but suggests that $W_{b\lambda} \to 0$ as $\lambda \to 0$. In other words, it predicts ever increasing levels of emission as we progress to shorter wavelengths. This enormous overestimate of the emission of short wavelengths became known as the "ultraviolet catastrophe".

The best theoretical model is due to Planck, who in the early 20th century used the developing ideas of quantisation of energy to derive an expression for $W_{b\lambda}$:

$$W_{b\lambda} = \frac{2\pi hc^2}{\lambda^5} \left(\frac{1}{\exp\left(\frac{hc}{\lambda kT}\right) - 1} \right) \qquad (1.13)$$

where h is Planck's constant. This bears close similarities with the Rayleigh-Jeans Law (Equation 1.12), but replaces classical equipartition of energy (in terms of kT) with quantised energy packets (photons) of energy $h\nu$ ($=hc/\lambda$). Equation 1.13 fits experimental data extremely well, and proved to be one of the early, great successes of the new quantum theory of radiation.

Planck's Law can be simplified at certain mathematical limits:

(i) In the limit of small wavelength and/or low temperature, $hc/\lambda kT$ and, hence, $\exp(hc/\lambda kT)$ will be much greater than 1, in which case Equation 1.13 reduces to:

$$W_{b\lambda} = \frac{2\pi hc^2}{\lambda^5} \exp\left(-\frac{hc}{\lambda kT} \right) \qquad (1.14)$$

This expression is known as Wien's Approximation, and may be used, to engineering accuracy, for visible and shorter wavelengths up to around 3000 K.

(ii) In the limit of long wavelength and/or high temperature, $hc/\lambda kT$ will be small and can be expanded in a Taylor's series approximation, whereby $e^x \approx 1 + x$ for small x. Applying this to Equation 1.13 gives

$$W_{b\lambda} = \frac{2\pi ckT}{\lambda^4}$$

which is exactly the same as Rayleigh-Jeans Law (Equation 1.12), and hence validates use of that equation at long wavelengths.

1.6   Stefan–Boltzmann Law

Before Planck made his theoretical breakthrough, Stefan and Boltzmann had independently used classical thermodynamics to show that the total emission (W_b) from a surface varies with the fourth power of absolute temperature. This result can also be derived from Planck's Law by substituting Equation 1.13 into Equation 1.11 and integrating:

$$W_b = \int_0^\infty W_{b\lambda}\,d\lambda = 2\pi hc^2 \int_0^\infty \frac{d\lambda}{\lambda^5\left(\exp\left(\frac{hc}{\lambda kT}\right) - 1\right)}$$

It is convenient to introduce the transformation

$$y = \frac{hc}{\lambda kT} \qquad \Rightarrow \qquad d\lambda = -\frac{hc}{kTy^2}\,dy$$

from which we obtain, after rearrangement:

$$W_b = \frac{2\pi k^4 T^4}{h^3 c^2} \int_0^\infty \frac{y^3}{e^y - 1}\,dy$$

The integral is a standard form, equal to ($\pi^4/15$), which gives:

$$W_b = \frac{2\pi^5 k^4 T^4}{15 h^3 c^2} = \sigma T^4 \qquad (1.15)$$

All terms on the right-hand side (except T) are fundamental constants, and are combined to give the Stefan–Boltzmann constant (σ), which is equal to 5.67×10^{-8} W/m^{2}K^4. This equation is one of the governing equations of radiation heat transfer and forms the basis of all heat transfer calculations.

1.7   Wien's displacement law

In Figure 1.4 it was noted that the maximum in $W_{b\lambda}$ moves to lower wavelengths as the temperature increases. Wien's Displacement Law (not to be confused with Wien's Approximation) relates the wavelength λ_{max} at which this maximum occurs to the temperature of the emitting body, and can be derived by differentiation of Planck's Law. We have

$$W_{b\lambda} = \frac{2\pi hc^2}{\lambda^5}\left(\frac{1}{\exp\left(\frac{hc}{\lambda kT}\right) - 1}\right) \qquad (1.13)$$

$$\frac{dW_{b\lambda}}{d\lambda} = -\frac{10\pi hc^2}{\lambda^6}\left(\frac{1}{\exp\left(\frac{hc}{\lambda kT}\right) - 1}\right) + \frac{2\pi hc^2}{\lambda^5}\left(\frac{1}{\exp\left(\frac{hc}{\lambda kT}\right) - 1}\right)^2 \frac{hc}{\lambda^2 kT}\exp\left(\frac{hc}{\lambda kT}\right)$$

Setting this equal to zero at $\lambda = \lambda_{max}$ gives after rearrangement:

$$5(e^x - 1) = xe^x \qquad \text{where } x = \frac{hc}{\lambda_{max}kT}.$$

This can be solved numerically to give $x = 4.965$, which leads to Wien's Law:

$$\lambda_{max} = \frac{hc}{4.965kT} = \frac{2.898 \times 10^{-3}}{T} \qquad (1.16)$$

Values of λ_{max} for various temperatures are given in Table 1.1. At ambient temperatures all radiation is in the infra-red part of the spectrum ($\lambda_{max} > 750$ nm), and no radiation is visible to the eye. As the temperature increases, radiation starts to be emitted in the red part of the visible region. The maximum continues to be in the infra-red, even at so-called "white heat". Only at temperatures of a few thousand Kelvin does the maximum move to the visible region (hence the sun appears yellow); at temperatures over 15 000 K the maximum is in the ultra-violet and surfaces emit more blue light than red (hence the hottest stars appear blue).

Table 1.1 λ_{max} for various temperatures calculated from Wien's Law

Temperature (K)	λ_{max}(m)	Surface Appearance
300	10^{-5}	Not visible
1 000	2.9×10^{-6}	Dull Red
1 500	1.9×10^{-6}	Orange-red
2 000	1.5×10^{-6}	"White" heat
6 000	4.8×10^{-7}	Yellow stars
30 000	10^{-8}	Blue stars

1.8 Total and monochromatic absorptivity and emissivity

We have so far established that real surfaces both emit and absorb less than a black body at the same temperature, and that the monochromatic emissivity and absorptivity are spectrally dependent. Equation 1.10 links the black body spectrum with the monochromatic and total emissivities. Since absorptivity is the fraction of incident radiation absorbed, we may also write:

$$\alpha = \frac{\int\limits_0^\infty \alpha_\lambda G_\lambda d\lambda}{\int\limits_0^\infty G_\lambda d\lambda} \qquad (1.17)$$

There is an important link between absorptivity and emissivity, known as Kirchhoff's Law, which is of great use in practical systems. Consider a small, non-black body contained within a black, isothermal enclosure (Figure 1.5). Furthermore, let us allow the body (surface 1) to come to thermal equilibrium with the walls of the enclosure (surface 2). This means that the temperature of the walls and the body are equal, and that the incident flux of every wavelength equals the leaving flux (radiosity) for that wavelength from the

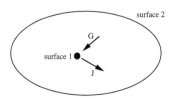

Figure 1.5 Radiation exchange between a small body and a large enclosure.

surface. From the definition of radiosity (Equation 1.6), we may write for surface 1:

$$G_{\lambda 1} = J_{\lambda 1} = W_{\lambda 1} + \rho_\lambda G_{\lambda 1}$$

where $G_{\lambda 1}$, $J_{\lambda 1}$ and $W_{\lambda 1}$ are the monochromatic incident flux, radiosity, and emissive flux, respectively, for surface 1. Since all radiation incident on surface 1 comes from surface 2 (which is black), $G_{\lambda 1}$ must equal $W_{b\lambda 2}$. Using Equations 1.5 and 1.8, we obtain:

$$W_{b\lambda 2} = \varepsilon_\lambda W_{b\lambda 1} + (1 - \alpha_\lambda)W_{b\lambda 2}$$

$$\Rightarrow \qquad \alpha_\lambda W_{b\lambda 2} = \varepsilon_\lambda W_{b\lambda 1}$$

As the temperatures of the two surfaces are equal, then $W_{b\lambda 1}=W_{b\lambda 2}$, since the monochromatic emissive power depends only on temperature. Thus we get Kirchhoff's Law:

$$\alpha_\lambda = \varepsilon_\lambda \qquad\qquad (1.18)$$

Both α_λ and ε_λ depend *only* on the nature and properties of surface 1 and are independent of surface 2. Therefore, although this expression has been derived for equilibrium conditions, it holds for *all* conditions, i.e. surface 2 need not be black and thermal equilibrium is not necessary. (N.B. A more rigorous treatment shows that the above expression holds only for a diffuse surface, i.e. where there is no directional dependence of α_λ and ε_λ.)

The above treatment can be extended to cover total absorptivity and emissivity: an energy balance requires the total radiosity (J) to equal the total incident flux (G), leading to the equality $\alpha = \varepsilon$ at thermal equilibrium. We *cannot*, however, now generalise this equality to non-equilibrium cases. Recalling Equations 1.10 and 1.17 for total emissivity and absorptivity, we have for surface 1:

$$\varepsilon = \frac{\int_0^\infty \varepsilon_\lambda W_{b\lambda 1}\,d\lambda}{\int_0^\infty W_{b\lambda 1}\,d\lambda} \qquad \text{and} \qquad \alpha = \frac{\int_0^\infty \alpha_\lambda G_{\lambda 1}\,d\lambda}{\int_0^\infty G_{\lambda 1}\,d\lambda}$$

The total emissivity is a function of surface 1 only, since both ε_λ and $W_{b\lambda 1}$ depend only on the properties of surface 1. But total absorptivity is a function of both surface 1 and surface 2, since α_λ is a function of surface 1 and $G_{\lambda 1}(= W_{b\lambda 2})$ is a function of the emission characteristics of surface 2. Thus in general $\alpha \neq \varepsilon$, except at thermal equilibrium. This restriction means that the equality appears to be of little value in heat transfer processes, which by their very nature are non-equilibrium systems. However, there is one further special case where the equality does hold.

Consider the case where both α_λ and ε_λ are constant and independent of wavelength. Both can then be taken outside the above integrals, allowing the integrals to be cancelled. We then have $\alpha = \alpha_\lambda$ and $\varepsilon = \varepsilon_\lambda$. Using Equation 1.18 (which is generally applicable) leads to $\alpha = \varepsilon$ under all conditions (i.e. non-equilibrium, non-black surfaces) where α_λ and ε_λ are constant. We shall see that using this equality is extremely useful in

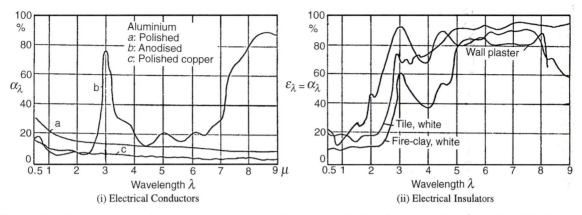

Figure 1.6 Monochromatic emissivity as a function of wavelength for a selection of metals and non-metals (from Kreith and Bohn).

engineering calculations, although the assumptions made in its derivation must always be remembered. Objects for which α_λ and ε_λ are independent of wavelength are generally called *grey* bodies, and any assumption that $\alpha = \varepsilon$ in non-equilibrium systems is referred to as the *grey body approximation*. The approximation is valid for most materials in the infra-red part of the spectrum ($\lambda = 1$ to $10\ \mu$m), but can be hopelessly inaccurate in the visible region. In engineering applications, the grey body appromixation can be used in, say, furnaces, where a large proportion of the heat transfer is due to infra-red heating; it should be used with extreme caution (preferably not at all) in solar heating systems.

There is a wealth of emissivity data available in the literature, a brief sample of which is shown here. Figure 1.6 shows monochromatic emissivity data as a function of wavelength for a small selection of metals and non-metals. It is noticeable that polished metals have very low emissivities (hence a high reflectivity) and are almost grey, with little variation in ε_λ with wavelength. Non-metals show a wide variation in ε_λ: emissivities in the visible region are low, but materials are almost black in the infra-red region. Table 1.2 shows total emissivity as a function of temperature for some common metals and non-metals. Metals generally have low emissivities, which rise slowly with temperature, whereas non-metals have high emissivities at ambient temperatures, which decrease with temperature. Note that the condition of the surface, particularly the degree of oxidation, can have a very marked effect on emissivity. Many of the values are only approximate and are averages of values taken from a number of different sources.

1.9 Black body temperature

We have established that the emissive power from a real surface of temperature T_s and emissivity ε is less than that from a black surface at the same temperature. We can imagine that there will exist a black surface at some

Table 1.2   Total emissivity as a function of temperature for a range of materials

	25 °C	200 °C	500 °C	1000 °C
Aluminium (polished)	0.04	0.05	0.07	0.15
Aluminium (oxidised)	0.20	0.22	0.33	
Brass (polished)	0.03	0.03	0.03	
Brass (oxidised)	0.45	0.53	0.75	
Copper (polished)	0.02	0.02	0.04	0.06
Copper (oxidised)	0.85	0.83	0.75	
Steel (polished)	0.07	0.08	0.10	0.20
Alumina		0.70	0.45	0.20
Brick	0.90	0.90	0.70	0.40
Carbon black	0.95	0.97	0.99	0.99
Glass	0.90	0.90	0.70	0.50
Graphite (polished)	0.45	0.95	0.95	
Limestone	0.95	0.83	0.75	
Paper	0.95	0.85	0.70	0.50
White Pigments	0.90	0.80	0.60	0.40

lower temperature (T_b) whose emissive power is the same as the real surface at T_s. We can therefore write:

$$W = \varepsilon \sigma T_s^4 = \sigma T_b^4$$

$$T_b = T_s \varepsilon^{0.25} \tag{1.19}$$

T_b is known as the *black body temperature* of the real surface and is always less than the actual temperature. Note that we are equating the total emissive fluxes; the spectral distribution of the real surface at T_s and the black body at T_b will be different. We will see later that black body temperature is an important concept in the measurement of high temperatures, because many techniques determine the black body temperature, rather than the surface temperature.

2 View factors

2.1 Introduction

The previous chapter dealt with the emission and absorption characteristics of surfaces. We will now turn our attention to radiative transfer from one surface to another. A surface generally emits radiation in all directions simultaneously. If the receiving surface is located some distance away from the radiation source, then not all the radiation leaving the source will impinge on the receiving surface. As the surface moves towards the source, increasingly more radiation will be received. This is obvious in everyday life: the closer we are to a heat source, the hotter we feel. Clearly the source is not emitting more radiation as we get closer, but a larger proportion is landing on us. This leads to the concept of the *view factor* (also known as the *configuration factor* or *shape factor*) which takes the geometry of the system into account. At this stage we shall assume that the medium between surfaces does not absorb or emit radiation.

2.2 Definition and properties of view factors

Consider two black surfaces 1 and 2, both of which radiate in all directions. We can define the view factor F_{12} as the fraction of the radiation emitted by surface 1 which is directly intercepted by surface 2. Similarly, the view factor F_{21} is the fraction of the radiation emitted by surface 2 which is intercepted by surface 1. The rate of emission (in Watts) from surface 1 is $W_{b1}A_1$, so the rate at which radiation arrives at surface 2 is $Q_{12} = W_{b1}A_1F_{12}$. Similarly the rate at which emission from surface 2 arrives at surface 1 is $Q_{21} = W_{b2}A_2F_{21}$. At thermal equilibrium, the temperature of the two surfaces will be equal, the emissive fluxes W_{b1} and W_{b2} will be equal and there will be no net heat flow between the two surfaces. Therefore

$$Q_{12} = Q_{21}$$
$$\Rightarrow \quad A_1F_{12} = A_2F_{21} \tag{2.1}$$

Equation 2.1 is known as the Reciprocity Relation. Note that it is dependent only on the geometry of the system, so it holds for all systems, even though it has been derived for black surfaces at thermal equilibrium.

There is one important property of view factors, known as the Summation Relation. Consider an enclosure containing N surfaces. Since all the radiation emitted from surface 1 must land somewhere in the enclosure, we can write:

$$F_{11} + F_{12} + F_{13} + \ldots\ldots + F_{1N} = 1$$

or, more generally for any surface i:

$$\sum_{j=1}^{j=N} F_{ij} = 1 \tag{2.2}$$

Note that this expression introduces the notion of a "self-view factor" such as F_{11}. For a concave surface, some of the radiation emitted from that surface may be intercepted by another part of the same surface. For instance in Figure 1.5, all radiation emitted by surface 1 is received by surface 2, so $F_{11} = 0$ and $F_{12} = 1$; however, some of the radiation from surface 2 misses surface 1 and lands elsewhere on surface 2, so $F_{22} \neq 0$.

Clearly in order to carry out a heat transfer analysis, the numerical values of the various view factors are required. The remainder of this chapter is concerned with some of the methods available for their determination.

2.3   Tables and charts

The view factors for many geometries have been evaluated over the years and are given in a range of reference books, either in the form of an algebraic equation or a chart, from which the view factor may be determined. Figure 2.1 gives two examples of a view factor chart, where the view factor is plotted as a function of the geometry for two parallel plates (Figure 2.1a) and two perpendicular plates (Figure 2.1b). Major compilations of view factor algebra and charts are given by Siegel and Howell (*Thermal radiation heat transfer*) by Modest (*Radiative heat transfer*) and by Howell (*A catalog of radiation configuration factors*). If literature values are not available, then view factors must be calculated by the user. A numerical example using charts is given in Section 2.8.

(a)
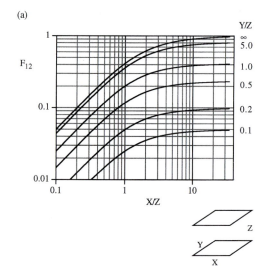

(b)
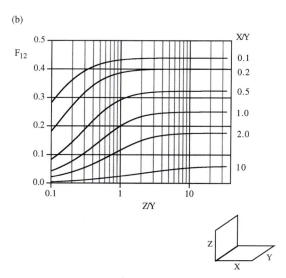

Figure 2.1   View Factor charts for (a) two parallel, rectangular plates of equal size, (b) two perpendicular plates with a common edge.

2.4   Cosine Law

Consider two area elements dA_1 and dA_2 (Figure 2.2), which may be part of two larger, finite areas A_1 and A_2. Let the line connecting the elements be of length r, and the angles between that line and the normals to the elements be θ_1 and θ_2 respectively. Assuming that the emitted radiation is isotropic, the rate of heat transfer (dQ_{12}) between the two elements can be derived purely from the geometry. It will be proportional to:

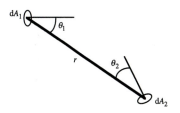

Figure 2.2   Radiative transfer between two area elements.

(i)   the apparent area of dA_2 as seen from dA_1 ($= dA_2 \cos \theta_2$);
(ii)   the apparent area of dA_1 as seen from dA_2 ($= dA_1 \cos \theta_1$);
(iii) the inverse square of the distance between the elements ($= 1/r^2$)

We can therefore write

$$dQ_{12} = \frac{I(dA_1 \cos \theta_1)(dA_2 \cos \theta_2)}{r^2} \qquad (2.3)$$

where I is a constant of proportionality. For finite areas this can be integrated over each surface to give

$$Q_{12} = \int\limits_{A_1} \int\limits_{A_2} \frac{I \cos \theta_1 \cos \theta_2 \; dA_1 dA_2}{r^2} \qquad (2.4)$$

We can determine I by evaluating this integral for a system in which Q_{12} can be found using an alternative method. Consider a hemisphere of radius r, whose base is centred at dA_1 (Figure 2.3). All radiation from dA_1 will land on the curved surface of the hemisphere. A thin circular strip element (dA_2) on the surface subtends an angle $d\theta_1$ at dA_1 and will have thickness $rd\theta_1$ and radius $r\sin\theta_1$. Thus:

$$dA_2 = 2\pi r\sin \theta_1 \times rd\theta_1$$

Substituting into Equation 2.3, noting that $\theta_2 = 0$, since the hemisphere radius is normal to dA_2, gives:

$$dQ_{12} = 2\pi I \; dA_1 \cos \theta_1 \sin \theta_1 \; d\theta_1$$

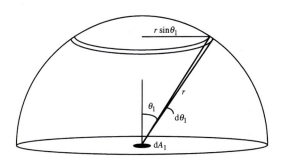

Figure 2.3   Radiative transfer between differential plane and ring elements.

This can be integrated over the curved surface of the hemisphere with respect to θ_1 to give the heat flow from dA_1 to the hemisphere:

$$Q_{12} = 2\pi I dA_1 \int_0^{\frac{\pi}{2}} \cos \theta_1 \sin \theta_1 d\theta_1 = 2\pi I \, dA_1 \left[\tfrac{1}{2} \sin^2 \theta_1 \right]_0^{\frac{\pi}{2}}$$

$$\Rightarrow \qquad Q_{12} = \pi I dA_1$$

Since *all* radiation from dA_1 must pass through the curved surface of the hemisphere, this equation gives us the total rate of heat flow from dA_1. This total rate of heat flow is also given by

$$Q_{12} = J_1 dA_1$$

where J_1 is the radiosity of dA_1. So I is simply J_1/π and Equation 2.4 may be rewritten

$$Q_{12} = \int_{A_1} \int_{A_2} \frac{J_1 \cos \theta_1 \cos \theta_2 \, dA_1 dA_2}{\pi r^2} \tag{2.5}$$

Recalling that $Q_{12} = J_1 A_1 F_{12}$ for exchange between any two finite areas, we have the following general expression for the view factor:

$$A_1 F_{12} = \int_{A_1} \int_{A_2} \frac{\cos \theta_1 \cos \theta_2 dA_1 dA_2}{\pi r^2} \tag{2.6}$$

This is known as the Cosine Law and gives us a method for calculating any view factor where the geometry is defined. Note that the symmetry of the integrand demonstrates the validity of the Reciprocity Relation for all systems (no restrictions to equilibrium or black surfaces have been used here): subscripts 1 and 2 are wholly interchangeable, such that $A_1 F_{12} = A_2 F_{21}$. Unfortunately the integral is rarely straightforward to evaluate. The double area integral required for exchange between finite areas requires a four-fold integration for many geometries. Although the development of computational techniques in recent years has helped greatly, analytical methods are often not practicable. Consequently other methods for determining view factors are frequently used.

Example 2.1.
Determine the view factor to a disc A_2 of diameter D from an elemental area dA_1 situated a perpendicular distance R below the centre of the disc.

Solution.
Consider an annular element thickness dx and area dA_2 a distance x from the centre of the disc (Figure 2.4).
From the geometry:

$$dA_2 = 2\pi x \, dx \qquad r = \sqrt{R^2 + x^2}$$

$$\cos \theta_1 = \cos \theta_2 = \frac{R}{\sqrt{R^2 + x^2}}$$

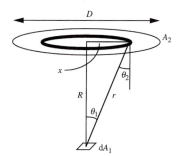

Figure 2.4 Geometry for Example 2.1.

Substituting these in Equation 2.6, noting that we require only a single integral because surface 1 is differential:

$$dA_1\, F_{12} = dA_1 \int_0^{\frac{D}{2}} \frac{2R^2 x dx}{(R^2 + x^2)^2} = \int_{x=0}^{x=\frac{D}{2}} \frac{R^2}{(R^2 + x^2)^2}\, d(x^2)$$

$$\Rightarrow \quad F_{12} = \frac{D^2}{4R^2 + D^2}$$

2.5   Unit Hemisphere Method

The Unit Hemisphere Method provides a way of using a geometric construction to determine one of the area integrals in Equation 2.6. Consider the two elemental areas dA_1 and dA_2 a distance r apart, and imagine a hemisphere of unit radius centred on dA_1 (Figure 2.5).

We now carry out a double projection of dA_2. First we project dA_2 onto the curved surface of the hemisphere, to give a shape of area dA_2'; then we project dA_2' onto the base of the hemisphere to give area dA_2''. Now consider the areas of each projection:

$$dA_2' = dA_2 \cos\theta_2 \times \frac{1}{r^2}$$

$$dA_2'' = dA_2' \cos\theta_1 = \frac{dA_2 \cos\theta_1 \cos\theta_2}{r^2}$$

The area of the base of a unit hemisphere is π, so

$$\text{Fraction of base occupied by } dA_2'' = \frac{dA_2''}{\pi} = \frac{dA_2 \cos\theta_1 \cos\theta_2}{\pi r^2}$$

If dA_2 is part of a finite area A_2 which forms a double projection of area A_2'' on the base of the hemisphere, then

$$\text{Fraction of base occupied by } A_2'' = \frac{A_2''}{\pi} = \int_{A_2} \frac{dA_2 \cos\theta_1 \cos\theta_2}{\pi r^2} \qquad (2.7)$$

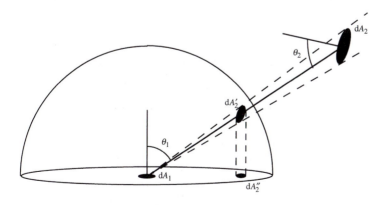

Figure 2.5   The Unit Hemisphere Method for evaluating the View Factor integral.

If we define ξ to be this fraction, then substitution into the Cosine Law (Equation 2.6) leads to

$$A_1 F_{12} = \int \xi \, dA_1 \qquad (2.8)$$

If this integral can be evaluated, then we can determine F_{12}. Note that if A_1 is elemental, then no integration is required, and $F_{12} = \xi$.

Example 2.2.
Determine the view factor from a disc of radius ρ to a sphere of radius r whose centre is located a perpendicular distance h above the centre of the disc.

Solution.
Consider an area element dA_1 on the disc a distance x from the centre of the disc (Figure 2.6), and imagine that dA_1 is located at centre of the base of a unit hemisphere. Let R be the beam length from dA_1 to the centre of the sphere (A_2), and let θ be the angle between the beam and the disc. From the geometry:

$$R = \sqrt{x^2 + h^2} \qquad \sin \theta = \frac{h}{R} \qquad (2.9)$$

The sphere A_2 will project onto the surface of the hemisphere as a circle of radius r/R (using the properties of similar triangles). This circle will project onto the base of the hemisphere as an ellipse with axes r/R and $r \sin\theta/R$. Therefore:

$$\text{Area of ellipse } (A_2'') = \frac{\pi r^2 \sin \theta}{R^2}$$

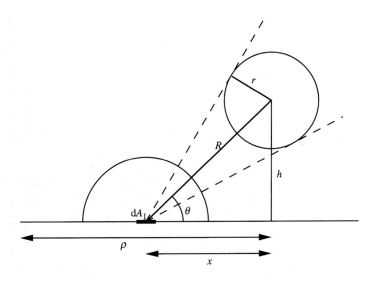

Figure 2.6 Geometry for Example 2.2.

Using Equations 2.7 and 2.9:

$$\xi = \text{Fraction of base occupied} = \frac{A_2''}{\pi} = \frac{r^2 \sin\theta}{R^2} = \frac{r^2 h}{(x^2 + h^2)^{\frac{3}{2}}}$$

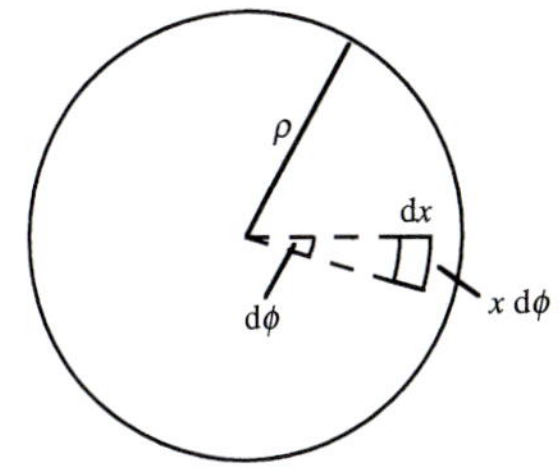

Suppose that element dA_1 on the disc has dimensions $dx \times x\,d\phi$, where $d\phi$ is the angle subtended by the element at the centre of the disc (Figure 2.7). Substituting for ξ and dA_1 into Equation 2.8 gives:

$$A_1 F_{12} = \pi\rho^2 F_{12} = \int\limits_{\phi=0}^{\phi=2\pi} \int\limits_{x=0}^{x=\rho} \frac{r^2 hx\,dx\,d\phi}{(x^2+h^2)^{\frac{3}{2}}}$$

which may be evaluated to give:

$$F_{12} = \frac{2r^2}{\rho^2}\left(1 - \frac{h}{\sqrt{\rho^2 + h^2}}\right)$$

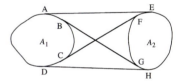

Figure 2.7   Relation between area element dA_1 and the whole disc.

2.6   Hottel's crossed-strings method

Hottel's method can be used in systems which are essentially two-dimensional, i.e. where the surfaces are of constant cross-section, constant separation, and their length is much greater than their separation. Although this would appear to limit its applicability severely, it can be used in many engineering situations, such as long furnaces.

Consider two parallel objects of infinite length, and imagine two strings wrapped round a cross-section (Figure 2.8): let the first be a single loop, which will be tangential to the surfaces at A, D, E, and H; let the second be a figure-of-eight, which will be tangential at B, C, F, and G. Hottel's Method says that the view factor from surface 1 to surface 2 is given by:

$$F_{12} = \frac{1}{2P_1}\left[\,(\overline{ABGH} + \overline{DCFE}) - (\overline{AE} + \overline{DH})\,\right] \qquad (2.10)$$

where P_1 is the perimeter of surface 1 and $\overline{ABGH}$ represents the length of string ABGH. In words, we have

$$F_{12} = \frac{1}{2P_1}\,[(\text{sum of the crossed strong lengths}) - $$
$$(\text{sum of the uncrossed string lengths})]$$

Figure 2.8   Hottel's Crossed-Strings Method for two general surfaces.

This equation may be derived by considering the geometry of the system. First, note that radiation from only the portion ABCD on surface 1 can be intercepted by surface 2, and only the portion EFGH on surface 2 can receive that radiation. Now consider the "triangle" BDH, and suppose that it forms an imaginary enclosure where the side lengths BD, BH, and DH are *a*, *b*, and *c*, respectively. We can further imagine view factors between the three sides of the triangle. From the properties of view factors:

$$F_{aa} = F_{bb} = F_{cc} = 0 \qquad \text{(all sides are convex)}$$
$$F_{ab} + F_{ac} = 1; \quad F_{ba} + F_{bc} = 1; \quad F_{ca} + F_{cb} = 1 \qquad \text{(Summation)}$$

$$aF_{ab} = bF_{ba}; \quad aF_{ac} = cF_{ca}; \quad bF_{bc} = cF_{cb} \qquad \text{(Reciprocity)}$$

These can be rearranged to give an equation for each view factor in terms only of the side lengths:

$$F_{ab} = \frac{a + b - c}{2a} \qquad F_{ba} = \frac{a + b - c}{2b} \qquad F_{ca} = \frac{a + c - b}{2c}$$

$$F_{ac} = \frac{a + c - b}{2a} \qquad F_{bc} = \frac{c + b - a}{2b} \qquad F_{cb} = \frac{c + b - a}{2c} \qquad (2.11)$$

Now, the radiation leaving surface 1 from the section BD $= \overline{BD}\, J_1$. So, the radiation leaving BD which crosses DH, and hence misses surface 2 $= \overline{BD} J_1 F_{ac}$. Since no radiation leaving AB crosses DH (as it is out of line-of-sight), this also represents the radiation leaving ABCD which crosses DH. Substituting for F_{ac} from Equations 2.11 gives:

$$\text{Radiation from ABCD which crosses DH} = \tfrac{1}{2} J_1 \big(\overline{ABCD} + \overline{DH} - \overline{ABGH} \big)$$

$$(2.12)$$

A similar analysis can be carried out on the "triangle" ACE, from which we find:

$$\text{Radiation from ABCD which crosses AE} = \tfrac{1}{2} J_1 \big(\overline{ABCD} + \overline{AE} - \overline{DCFE} \big)$$

$$(2.13)$$

Note also that

$$\text{Radiation from ABCD in all directions} = J_1 \overline{ABCD} \qquad (2.14)$$

All radiation from ABCD must *either* cross DH, *or* cross AE, *or* impinge on surface 2. So combining Equations 2.12 to 2.14 gives:

Radiation from ABCD impinging on surface $2 = Q_{12}$

$$= J_1 \overline{ABCD} - \tfrac{1}{2} J_1 \big(\overline{ABCD} + \overline{DH} - \overline{ABGH} \big) - \tfrac{1}{2} J_1 \big(\overline{ABCD} + \overline{AE} - \overline{DCFE} \big)$$

$$= \tfrac{1}{2} J_1 \big[\big(\overline{ABGH} + \overline{DCFE} \big) - \big(\overline{AE} + \overline{DH} \big) \big] \qquad (2.15)$$

Using the definition of a view factor, we have an alternative expression for Q_{12}, viz.:

$$Q_{12} = P_1 J_1 F_{12} \qquad (2.16)$$

Combining Equations 2.15 and 2.16 gives Equation 2.10:

$$F_{12} = \frac{1}{2P_1} \big[\big(\overline{ABGH} + \overline{DCFE} \big) - \big(\overline{AE} + \overline{DH} \big) \big] \qquad (2.10)$$

Example 2.3.
Determine the view factor between two long, parallel tubes of diameter D and centre-to-centre spacing L.

Solution.
Crossed and uncrossed strings are drawn and labelled in Figure 2.9, where θ is defined as the angle $\angle$AOE, which, using simple geometry, also equals $\angle$OFE.

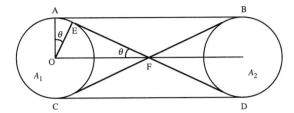

Figure 2.9 Geometry for Example 2.3.

Hottel's Method tells us:

$$F_{12} = \frac{1}{2P_1}[(\text{Sum of the crossed string lengths})$$
$$- (\text{Sum of the uncrossed string lengths})]$$
$$= \frac{1}{2P_1}[(\overline{AD} + \overline{BC}) - (\overline{AB} + \overline{CD})] \qquad (2.17)$$

From Figure 2.9:

$$P_1 = \pi D \qquad \overline{AB} = \overline{CD} = L \qquad (2.18)$$

$$\overline{AD} = \overline{BC} = 2(\overline{AE} + \overline{EF}) = 2\left(\frac{D}{2}\theta + \sqrt{\left(\frac{L}{2}\right)^2 - \left(\frac{D}{2}\right)^2}\right) \qquad (2.19)$$

From the triangle OEF, $\sin\theta = (L/D)$, so $\theta = \sin^{-1}(L/D)$. Substituting into Equation 2.19 gives:

$$\overline{AD} = \overline{BC} = D\sin^{-1}\left(\frac{D}{L}\right) + \sqrt{L^2 - D^2}$$

This together with Equations 2.18 can be substituted into Equation 2.17 to give (after rearrangement):

$$F_{12} = \frac{1}{\pi}\left[\sin^{-1}\left(\frac{1}{X}\right) + \sqrt{X^2 - 1} - X\right]$$

where $X = L/D$.

2.7 Monte Carlo methods

For complicated geometries where there are no published data and direct calculation is not possible, numerical techniques are required. Numerical integration of the view factor integrals is possible using computers, but perhaps the most common method is the Monte Carlo Method. The basis of the method for view factor determination is to select at random a series of beams of radiation from surface 1, and to determine the proportion which impinge on surface 2. The procedure may be summarised as follows:

(1) Define the geometry for the two surfaces of interest.
(2) Select at random a point on surface 1.
(3) Select at random a direction for a beam of radiation leaving surface 1 at that point.
(4) Determine whether that beam impinges on surface 2.
(5) Repeat steps (2) to (4) for a large number of beams.
(6) The view factor is the ratio of the number of beams impinging on surface 2 to the total number.

The greater the number of beams used, the greater the accuracy, i.e. the closer the value obtained is to the real value. Depending on the size and complexity of the surfaces, it can require several hundred or thousand beams to give a view factor close to the real value, which means that this method is practicable only using computer techniques. Each step therefore has to be put into a suitable algebraic form. The technique will be illustrated here for a simple system in two dimensions, viz. two parallel plates, whose length is much greater than the width. A much more comprehensive treatment of Monte Carlo methods in two and three dimensions is given in the major textbooks.

Define two parallel plates (see Figure 2.10), with surface 1 having ends at coordinates (x_1,y_1) and (x_2,y_2), and surface 2 having ends at (x_3,y_3) and (x_4,y_4). Choice of origin and coordinate system is entirely arbitrary, so we let surface 1 lie along the x-axis ($y_1 = y_2 = 0$), with its left-hand end at the origin ($x_1 = 0$). Consequently surface 2 lies parallel to the x-axis ($y_3 = y_4$). We can pick any point on surface 1 at random by using a random number generator to give a number R_1 between 0 and 1, which represents the fractional distance along the surface from x_1. The selected point x' is related algebraically to R_1:

$$R_1 = \frac{x' - x_1}{x_2 - x_1}$$

$$\Rightarrow \quad x' = x_1 + R_1(x_2 - x_1)$$

We must now choose a direction in which the beam leaves the surface. Let us define θ as the angle between the normal to surface 1 at x' and any beam of radiation from that point, which means that θ can vary between $-\pi/2$ and $+\pi/2$. We can pick any angle from surface 1 by using the random number

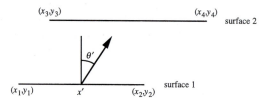

Figure 2.10 Arrangement for two parallel plates, showing a beam of radiation from a random point on surface 1 in a random direction.

generator to give a number R_2, which represents where the beam is between $-\pi/2$ and $+\pi/2$. Now, if the intensity of radiation normal to surface 1 is I, then the energy emitted in direction θ will be $I\cos\theta$. The selected angle θ' is related to R_2 by:

$$R_2 = \frac{\int\limits_{-\frac{\pi}{2}}^{\theta'} I\cos\theta d\theta}{\int\limits_{-\frac{\pi}{2}}^{+\frac{\pi}{2}} I\cos\theta d\theta} = \frac{1}{2}(\sin\theta' + 1)$$

$$\Rightarrow \quad \theta' = \sin^{-1}(2R_2 + 1)$$

We now have a position x' on surface 1 and a beam direction θ', and must now determine whether the beam hits surface 2. To do this we construct algebraic equations for the beam and for surface 2, and then test where the two lines intercept. We have:

$$\text{Equation of Beam:} \quad y = \frac{x - x'}{\tan\theta'} \qquad (2.20)$$

Surface 2 is the part of the line $y = y_3$ between $x = x_3$ and $x = x_4$. Substituting $y = y_3$ into Equation 2.20 gives us a point x where the two lines intersect. If $x_3 \leq x \leq x_4$, then the beam does impinge on surface 2 and we record a hit; if $x < x_3$ or $x > x_4$, then the beam does not impinge on surface 2 and we record a miss. We then repeat this procedure for a very large number of randomly selected beams. Use of computers allows us to compute several thousand beams very quickly, and hence determine the view factor to a high degree of accuracy.

The method can be extended to three dimensions, but requires additional random numbers. For example, a finite plate in the xy plane requires two numbers to define a point (x', y'), one for each of the x and y coordinates. The beam direction requires two angles to define it, e.g. a circumferential angle and an angle of inclination, each of which requires a random number. Further details are given in the major textbooks.

2.8   View factor algebra

When embarking on view factor calculations, it is normally necessary to calculate every view factor from first principles or from charts. Depending on the geometry and the number of surfaces involved, we usually need to calculate only one or two view factors at most, and then determine the rest using the Reciprocity and Summation Relations (Equations 2.1 and 2.2). Consider a system of three surfaces, which will have nine view factors (F_{11}, F_{12}, F_{13}, F_{21}, F_{22}, F_{23}, F_{31}, F_{32}, F_{33}). These may be linked using reciprocity and summation:

$$F_{11} + F_{12} + F_{13} = 1$$
$$F_{21} + F_{22} + F_{23} = 1$$

$$F_{31} + F_{32} + F_{33} = 1$$

$$A_1 F_{12} = A_2 F_{21} \qquad A_1 F_{13} = A_3 F_{31} \qquad A_2 F_{23} = A_3 F_{32}$$

Assuming that the geometry is defined, i.e. all areas are known, then we have nine unknown view factors and six equations. So we need to determine only three view factors in order to solve for all the rest. In many systems there will be plane or convex surfaces where the self-view factor will be zero, so the problem quickly reduces itself to the determination from first principles of, perhaps, just one view factor.

Example 2.4.

Consider a rectangular box furnace with side length 5 m × 4 m × 3 m (Figure 2.11). The roof of the furnace can be taken to be surface 1, the floor surface 2, and the four walls can be treated as a *single* surface or area A_3. Determine all the view factors.

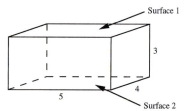

Figure 2.11 Geometry for Example 2.4.

Solution.

This is an example of the three-surface problem described above, so we need to determine three view factors and then use reciprocity and summation to solve for the rest. By inspection, F_{11} and F_{22} are both zero, because surfaces 1 and 2 are both planar. (F_{33} is not zero, as radiation can be transmitted from one wall to another.) We therefore need to determine just one view factor from first principles. Surfaces 1 and 2 are parallel plates of the same geometry as Figure 2.1a. Using the nomenclature of Figure 2.1, we have $X/Z = 1.67$, $Y/Z = 1.33$, leading to $F_{12} = 0.32$. The Summation Relation for surface 1 then gives $F_{13} = 0.68$. The symmetry of the system requires that F_{21} and F_{23} are also 0.32 and 0.68, respectively. Nothing that $A_1 = 20 \text{ m}^2$ and $A_3 = 54 \text{ m}^2$, we can use the Reciprocity Relation between surfaces 1 and 3 to give $F_{31} = 0.25$. Symmetry requires that F_{32} is also 0.25, and the Summation Relation gives $F_{33} = 0.50$.

View factors for some more complex arrangements can be found using the properties of reciprocity and summation. For example, suppose that we wish to determine the view factor F_{12} in Figure 2.12. Surfaces 1 and 2 are part of perpendicular planes, but they do not share a common edge, so Figure 2.1b cannot be applied directly. However, we can make use of the chart by noting that surfaces 1 and 2 are part of two planes which *are* perpendicular with a common edge (as shown with dashed lines in Figure 2.12). The chart will give view factors for $F_{(1+3)(2+4)}$, $F_{(1+3)4}$, $F_{3(2+4)}$, and F_{34}. The properties of view factors then give:

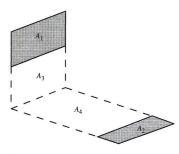

Figure 2.12 Determination of view factor for non-touching perpendicular planes.

$$F_{(1+3)2} = F_{(1+3)(2+4)} - F_{(1+3)4}$$

$$F_{32} = F_{3(2+4)} - F_{34}$$

$$F_{12} = F_{(1+3)2} - F_{32}$$

3   Heat exchange between black surfaces

3.1   Introduction

Various methods for determining view factors between surfaces were discussed in the previous chapter. We can now start solving problems of heat transfer between surfaces. In this chapter we will confine ourselves solely to radiation exchange between black surfaces, in which case the only radiation leaving a given surface is thermal emission. All incident radiation on a black surface is absorbed and none is reflected. Further, we shall assume that the medium between the surfaces neither emits nor absorbs radiation, i.e. it is non-participating. Non-black surfaces dealt with in Chapter 4, whereas participating media are considered in Chapter 5.

All radiative transfer problems involving surfaces are built around a consideration of the radiative fluxes incident on and leaving a given surface. Consider two black surfaces (1 and 2), where surface 1 is at a higher temperature than surface 2. The radiation from surface 1 to surface 2 will be the fraction of the emission from surface 1 which is incident on surface 2, i.e. $W_{b1}A_1F_{12}$. Similarly, the radiation from surface 2 to surface 1 will be $W_{b2}A_2F_{21} = W_{b2}A_1F_{12}$ (using Reciprocity). The net transfer of radiation (Q_{12}) from surface 1 to surface 2 will be the difference between these terms, i.e.

$$Q_{12} = A_1F_{12}(W_{b1} - W_{b2}) = A_1F_{12}(\sigma T_1^4 - \sigma T_2^4) \qquad (3.1)$$

Clearly, if the geometry, the view factor, and the temperatures of the two surfaces are known, then the rate of heat transfer can be calculated. This general approach, of analysing a configuration in terms of transfer between various pairs of surfaces, forms the basis of all radiation calculations.

3.2   Three-surface enclosures—the refractory surface

We can generalise the above approach to a system of black surfaces, all of which are at different temperatures. Equation 3.1 can be applied to the radiation between any pair of surfaces. The direct radiation Q_{ij} from surface i to surface j will be given by

$$Q_{ij} = A_iF_{ij}(\sigma T_i^4 - \sigma T_j^4) \qquad (3.2)$$

and the net radiation Q_i leaving surface i in all directions will be:

$$Q_i = \sum_j A_iF_{ij}(\sigma T_i^4 - \sigma T_j^4) \qquad (3.3)$$

We see that knowledge of the geometry, all surface temperatures, and the view factors will enable the rate of heat transfer to be determined. (Note on nomenclature for Q: a double subscript, such as Q_{ij}, is taken to mean the radiative transfer from surface i to surface j; a single subscript, such as Q_i, is taken to mean the radiation received/emitted by surface i from/to *all* other surfaces.)

Enclosure problems are common to many industrial engineering applications, e.g. furnaces. In high temperature furnaces, we usually find surfaces acting as a heat source (such as radiant panels or a bed of burning fuel), surfaces acting as a heat sink (i.e. the material being heated) and the furnace walls. Furnace walls will normally be well insulated in order to reduce heat losses by conduction through the walls to the surroundings. A perfectly insulated surface will exchange radiation with other surfaces within the enclosure, but there will be zero conductive heat flow to the surroundings. If convective heat transfer between the furnace gases and the walls is also negligible, then at steady state the incident flux and the emissive flux will be equal (if the surface is black), and the net radiation leaving the furnace wall will be zero. The temperature of the surface will, essentially, float in such a way that the emissive flux matches the incident flux. Such a surface is variously called a *refractory* surface, a *reradiating* surface, or an *adiabatic* surface. If we can assume that any surfaces in a multi-surface problem are refractory, then it greatly simplifies the analysis.

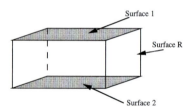

Figure 3.1 Schematic of a three-surface rectangular furnace.

Consider a rectangular box furnace (Figure 3.1) where all the surfaces are black, the whole of the roof (surface 1) is the heat source, the whole of the floor (surface 2) is the heat sink, and all the walls are refractory (surface R). Suppose that we wish to calculate the net rate of heat transfer to the heat sink, i.e. the material that we are interested in heating. We will assume that each of these three surfaces is at a constant temperature (T_1, T_2, and T_R, respectively). (This problem is, of course, somewhat idealised. The heat source and sink would rarely occupy the whole face of a furnace, and the temperatures would not be uniform, especially at the interfaces, where we would not expect discontinuities. However, this suits our purpose here.)

We approach the problem by thinking about the net radiative flux to surface 2. Surface 2 receives radiation from both surface 1 and surface R; it also emits radiation to surface 1 and surface R. In this example surface 2 is a plane, so there is no radiation received/emitted from/to elsewhere on surface 2. The net radiation received (Q_2) will be the difference between the incident and emitted radiation:

$$Q_2 = \text{Radiation received by surface 2} - \text{Radiation emitted by surface 2}$$
$$= (Q_{12} + Q_{R2}) - (Q_{21} + Q_{2R})$$
$$= A_1 F_{12} \sigma T_1^4 + A_R F_{R2} \sigma T_R^4 - A_2 F_{21} \sigma T_2^4 - A_2 F_{2R} \sigma T_2^4$$

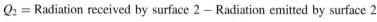

which on using the Reciprocity Relation gives:

$$Q_2 = A_1 F_{12} \left(\sigma T_1^4 - \sigma T_2^4 \right) + A_2 F_{2R} \left(\sigma T_R^4 - \sigma T_2^4 \right) \tag{3.4}$$

Equation 3.4 clearly shows the two contributions to the overall rate of heat transfer to surface 2: the first term represents direct exchange from the heat source, whereas the second term represents radiation received from the walls. If all the quantities on the right-hand side of the equation are known, then the rate of heat transfer may be calculated. The temperature of the walls (T_R) may be calculated, using the basic definition of a refractory surface:

Radiation received by surface R = Radiation emitted by surface R

$$Q_{1R} + Q_{2R} + Q_{RR} = Q_{R1} + Q_{R2} + Q_{RR}$$

$$A_1 F_{1R} \sigma T_1^4 + A_2 F_{2R} \sigma T_2^4 = A_R F_{R1} \sigma T_R^4 + A_R F_{R2} \sigma T_R^4$$

$$T_R^4 = \frac{A_1 F_{1R} T_1^4 + A_2 F_{2R} T_2^4}{A_R (F_{R1} + F_{R2})}$$

Using the Reciprocity Relation, we obtain

$$T_R^4 = \frac{F_{R1} T_1^4 + F_{R2} T_2^4}{F_{R1} + F_{R2}}$$

where the temperature of the walls is now expressed purely in terms of view factors and the temperatures of surfaces 1 and 2. Substitution of this into Equation 3.4 gives, after some rearrangement:

$$Q_2 = A_1 F_{12}(\sigma T_1^4 - \sigma T_2^4) + \frac{A_2 F_{2R} F_{R1}}{F_{R1} + F_{R2}} (\sigma T_1^4 - \sigma T_2^4)$$

Use of reciprocity then gives:

$$Q_2 = A_1 \left(F_{12} + \frac{F_{1R} F_{R2}}{F_{R1} + F_{R2}} \right)(\sigma T_1^4 - \sigma T_2^4) \tag{3.5}$$

Again, the two contributions to Q_2 are readily noticeable. It is worth noting at this point that because refractory surfaces behave as reradiators (incident flux = emitted flux), the radiation received by surface 2 from the walls can be thought of as radiation from surface 1 which arrives at surface 2 indirectly via the walls. Thus Equation 3.5 represents the sum of direct radiation (from 1 to 2) and indirect radiation (from 1 to 2 via the walls). (Some care must be exercised with this approach. The notion of indirect radiation must not be misconstrued into thinking of a refractory surface as a reflecting surface. Although the total incident flux and emitted flux from the refractory are equal, the spectral distributions of those fluxes will be very different, because the temperatures of the radiation sources are not the same. A black refractory surface is not a reflector—it absorbs radiation incident upon it, and emits radiation with the same total flux, but with a spectral distribution appropriate to its own surface temperature.) The bracketed combination of view factors in Equation 3.5 represents the fraction of radiation emitted by surface 1 which reaches surface 2 directly and indirectly with the assistance of the refractory surface. It can be thought of as a modified view factor, and is known as the *total radiation factor*. Equation 3.5 can therefore be rewritten:

$$Q_2 = A_1 \overline{F}_{12}(\sigma T_1^4 - \sigma T_2^4) \tag{3.6}$$

where $\overline{F}_{12}$ is the total radiation factor between surface 1 and surface 2 and is given by:

$$\overline{F}_{12} = F_{12} + \frac{F_{1R}F_{R2}}{F_{R1} + F_{R2}}$$

Clearly, $\overline{F}_{12}$ is always greater than F_{12}, since it includes the indirect path via the refractory. Total radiation factors have the same properties as view factors in that they obey both the summation and reciprocity relations.

3.3 Electrical Network Analogy

The method described in the previous section can, in principle, be used for any N-surface enclosure to develop a set of equations for solution. The complexity of the calculations increases as the number of surfaces increases. There is a method available which can reduce the number of calculations considerably, especially for grey surfaces (see Chapter 4). This has become known as the Electrical Network Analogy.

We have already seen that the radiative transfer (Q_{12}) between two black surfaces is given by:

$$Q_{12} = A_1 F_{12}(W_{b1} - W_{b2}) = A_1 F_{12}(\sigma T_1^4 - \sigma T_2^4) \qquad (3.1)$$

Closer inspection shows that the form of this equation is very similar to that for the flow of current I_{12} in an electrical circuit between nodes 1 and 2:

$$I_{12} = C(V_1 - V_2)$$

where C is the electrical conductance between points 1 and 2, and V_i is the electrical potential at node i. We can therefore construct an equivalent electrical network where each surface is represented by a node with a potential of σT^4, and the nodes are connected to each other by conductances $A_i F_{ij}$. The equivalence is shown diagrammatically in Figure 3.2.

In order to solve a radiation problem, we set up the equivalent electrical circuit and calculate the various unknowns using the rules of basic circuit theory, notably:

(1) conductances in parallel are additive;
(2) conductances in series add reciprocally;
(3) current flow into a node must equal current flow away from a node.

The last rule is effectively a statement of conservation of energy.

Figure 3.2 The equivalence of radiative transfer and an electrical circuit.

We can repeat the analysis of the three-surface black enclosure from Section 3.2 using the network approach. Figure 3.3 shows the equivalent circuit. Each of the three surfaces (1, 2, and R) is identified by a potential node, and the nodes are connected by the appropriate conductances. We require the net radiative transfer from surface 1 to surface 2, i.e. the net current from node 1 to node 2. In order to do this, we use the additive rules for conductances to replace the network of three conductances by a single conductance C_T. If the value of that single conductance can be determined, then calculation of the current becomes straightforward using:

$$Q_{12} = C_T(\sigma T_1^4 - \sigma T_2^4) \tag{3.7}$$

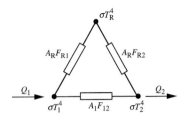

Figure 3.3 Equivalent electrical network for a three-surface enclosure (one heat source, one heat sink, one refractory surface).

The total conductance in Figure 3.3 between nodes 1 and 2 is given by:

$$C_T = C_D + C_R \tag{3.8}$$

where C_D is the "direct" conductance between nodes 1 and 2, and C_R is the "indirect" conductance via node R. C_R is itself a combination of two series conductances, such that:

$$\frac{1}{C_R} = \frac{1}{A_R F_{R1}} + \frac{1}{A_R F_{R2}}$$

$$\Rightarrow \quad C_R = \frac{A_R F_{R1} F_{R2}}{F_{R1} + F_{R2}}$$

Substitution into Equation 3.8 gives:

$$C_T = A_1 F_{12} + \frac{A_R F_{R1} F_{R2}}{F_{R1} + F_{R2}}$$

$$\Rightarrow \quad C_T = A_1 \left[F_{12} + \frac{F_{1R} F_{R2}}{F_{R1} + F_{R2}} \right] \qquad \text{(using reciprocity)}$$

This is the net conductance between nodes 1 and 2. Substitution into Equation 3.7 gives the rate of radiative transfer.

$$Q_{12} = A_1 \left[F_{12} + \frac{F_{1R} F_{R2}}{F_{R1} + F_{R2}} \right] (\sigma T_1^4 - \sigma T_2^4) = A_1 \overline{F}_{12}(\sigma T_1^4 - \sigma T_2^4)$$

This is exactly the same expression as before (Equation 3.5), but the derivation has required considerably less algebra. If we require the temperature of the refractory, we apply the third circuit rule to the refractory node:

Sum of currents into node R = Sum of currents away from node R

$$A_R F_{R1}(\sigma T_1^4 - \sigma T_R^4) = A_R F_{R2}(\sigma T_R^4 - \sigma T_2^4)$$

$$\Rightarrow \quad T_R^4 = \frac{F_{R1} T_1^4 + F_{R2} T_2^4}{F_{R1} + F_{R2}} \qquad \text{(as before)}$$

The network analogy has a certain elegance and is very appealing for these relatively straightforward systems. For more complicated systems involving more surfaces and hence more complicated equivalent circuitry, calculations can quickly become tedious, if not impossible to solve by hand. For multi-surface enclosures more sophisticated computational techniques are needed, because analysis of such systems usually leads to a large number of equations with a large number of unknowns. Fortunately matrix algebra can be used in conjunction with digital computers and this is recommended for all but the simplest problems. A discussion of matrix solutions for black and grey enclosures is given in Chapter 4.

4 Heat exchange between grey surfaces

4.1 Introduction

Any assumption of black body behaviour is an idealisation, which may be closely approximated by some surfaces, but is wholly inadequate for many practical systems. For non-black surfaces, we have the additional complication of reflection of incident radiation, such that the radiation leaving the surface is composed of emission plus reflection. We will confine ourselves in this book to opaque, grey surfaces, i.e. surfaces where we can assume that $\varepsilon = \alpha = 1 - \rho$. There are three important equations which will be used when dealing with grey surfaces. These are the equivalent of Equations 3.2 and 3.3 previously derived for black surfaces.

Consider two grey surfaces (i and j) and the radiative fluxes associated with one of those surfaces. The radiation from surface i to surface j will be the fraction of the radiation leaving surface i which is incident on surface j, i.e. $J_iA_iF_{ij}$, where J_i, is the radiosity of surface i and includes both emission and any incident radiation which is reflected. Similarly, the radiation from surface j to surface i will be $J_jA_jF_{ji} = J_jA_iF_{ij}$ (using reciprocity). The net transfer of radiation (Q_{ij}) directly from surface i to surface j will be the difference between these terms, i.e.

$$Q_{ij} = A_iF_{ij}(J_i - J_j) \tag{4.1}$$

Thus the transfer rate is dependent on the difference between the *radiosity* of each surface. This contrasts with black surfaces, where radiative transfer depends on the difference between the *emissive power* of each surface (Equation 3.2). The net radiation Q_i leaving surface i in all directions will be:

$$Q_i = \sum_j A_iF_{ij}(J_i - J_j) \tag{4.2}$$

There is an alternative formulation for Q_i. The net heat flow must be the difference between the leaving flux and the incident flux:

$$Q_i = (J_i - G_i)A_i \tag{4.3}$$

From the definition of radiosity (Equation 1.6), we have:

$$G_i = \frac{J_i - W_i}{\rho_i} = \frac{J_i - \varepsilon_i\sigma T_i^4}{\rho_i}$$

which on substituting into Equation 4.3 gives:

$$Q_i = A_i\left[\frac{\varepsilon_i\sigma T_i^4 - J_i(1 - \rho_i)}{\rho_i}\right]$$

Applying the grey body approximation ($\varepsilon = 1 - \rho$) then gives:

$$Q_i = \frac{A_i \varepsilon_i}{1 - \varepsilon_i} \left(\sigma T_i^4 - J_i \right) \tag{4.4}$$

Equations 4.1, 4.2, and 4.4 will form the basis of all radiative problems with grey surfaces.

4.2   Transfer between two surfaces

The additional complexities associated with non-black systems are illustrated in even the simplest system involving just two surfaces. The difficulties arise because of the radiosity terms in the governing equations. The temperature of a surface (and hence its emissive power) can be measured relatively easily, but the radiosity is almost certainly unknown. Therefore, analysis of any system involving non-black surfaces invariably requires solving a set of equations to obtain radiosities, from which heat transfer rates can then be determined.

Consider two very large parallel plates (Figure 4.1), with temperatures T_1, T_2 and emissivities ε_1, ε_2. Suppose that we wish to calculate the rate of radiative transfer from plate 1 to plate 2. Since all the radiation leaving one plate must be intercepted by the other plate, we have $G_1 = J_2$, $G_2 = J_1$, and $F_{12} = F_{21} = 1$. From Equation 4.1 the radiative flux is simply $(J_1 - J_2)$. The radiosities are unknown, so we use our basic definition (Equation 1.6) to link radiosities with the temperature and physical properties of each surface. Applying Equation 1.6 to each surface gives:

$$J_1 = \rho_1 G_1 + \varepsilon_1 \sigma T_1^4 = \rho_1 J_2 + \varepsilon_1 \sigma T_1^4$$

$$J_2 = \rho_2 G_2 + \varepsilon_2 \sigma T_2^4 = \rho_2 J_1 + \varepsilon_2 \sigma T_2^4$$

These can be rearranged as two simultaneous equations in J_1 and J_2:

$$J_1 - \rho_1 J_2 = \varepsilon_1 \sigma T_1^4 \tag{4.5}$$

$$-\rho_2 J_1 + J_2 = \varepsilon_2 \sigma T_2^4 \tag{4.6}$$

Eliminating J_2 from Equations 4.5 and 4.6 gives:

$$J_1 (1 - \rho_1 \rho_2) = \varepsilon_1 \sigma T_1^4 + \rho_1 \varepsilon_2 \sigma T_2^4 \tag{4.7}$$

Eliminating J_1 from Equations 4.5 and 4.6 gives:

$$J_2 (1 - \rho_1 \rho_2) = \rho_2 \varepsilon_1 \sigma T_1^4 + \varepsilon_2 \sigma T_2^4 \tag{4.8}$$

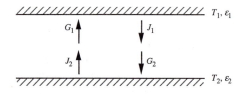

Figure 4.1   Geometry for radiative transfer between two large parallel plates.

Subtracting Equation 4.8 from Equation 4.7 and rearranging gives:

$$J_1 - J_2 = \frac{\varepsilon_1(1 - \rho_2)\sigma T_1^4 - \varepsilon_2(1 - \rho_1)\sigma T_2^4}{(1 - \rho_1\rho_2)} \qquad (4.9)$$

So far the analysis has been general to all non-black surfaces, and Q_{12} can be calculated if T, ε, and ρ are known for each surface. The expression can be simplified, if the surfaces are grey and opaque. If we introduce $\rho = 1 - \varepsilon$ for each surface, Equation 4.9 becomes, after some manipulation:

$$\text{Flux} = J_1 - J_2 = \left(\frac{1}{\varepsilon_1^{-1} + \varepsilon_2^{-1} - 1}\right)(\sigma T_1^4 - \sigma T_2^4) \qquad (4.10)$$

We have now achieved our aim and derived an expression for the radiative flux in terms of the temperature and physical properties (the emissivity) of the two surfaces. It is worth comparing Equation 4.10 with the flux for a pair of black parallel plates, which is simply $(\sigma T_1^4 - \sigma T_2^4)$. The first bracketed term in Equation 4.10 is a factor by which the radiative flux for black plates is reduced when the surfaces are grey. It is generally known as the *grey radiation factor*, and Equation 4.10 can be rewritten:

$$\text{Flux} = \mathcal{F}_{12}(\sigma T_1^4 - \sigma T_2^4) \qquad (4.11)$$

where $\mathcal{F}_{12}$ is the grey radiation factor for two large parallel plates. Since the emissivity must be between 0 and 1, $\mathcal{F}_{12}$ must be less than 1. In the more general case where $F_{12} < 1$, it can be shown that $\mathcal{F}_{12} \leq F_{12}$, with the equality holding for black surfaces. $\mathcal{F}_{12}$ can therefore be seen as a modified view factor from surface 1 to surface 2, which takes account of the non-black nature of the surfaces. In general we have

$$\text{Flux} = \mathcal{F}_{12}(\sigma T_1^4 - \sigma T_2^4) = F_{12}(J_1 - J_2)$$

and the flux is determined *either* from the emissive powers and the grey radiation factor, *or* from the radiosities and the view factor.

4.3 Network analogy for grey surfaces

We have seen how the solution of radiation problems for black surfaces can be accelerated by using an equivalent electrical circuit. The network approach can also be used for grey surfaces, provided that we take account of the non-ideal nature of the surface. Let us recall Equations 4.1 and 4.4:

$$Q_{ij} = A_i F_{ij}(J_i - J_j) \qquad (4.1)$$

$$Q_i = \frac{A_i \varepsilon_i}{1 - \varepsilon_i}(\sigma T_i^4 - J_i) \qquad (4.4)$$

Equation 4.1 suggests that direct radiative exchange between two surfaces can be represented as the current flow through a conductance connecting two radiosity nodes. Equation 4.4 suggests that the heat flow to/from a surface may be represented as the current flow through a conductance $A_i \varepsilon_i/(1 - \varepsilon_i)$ connecting a node at a potential equal to the black body emissive power for that surface to a radiosity node for that surface. This second term can be

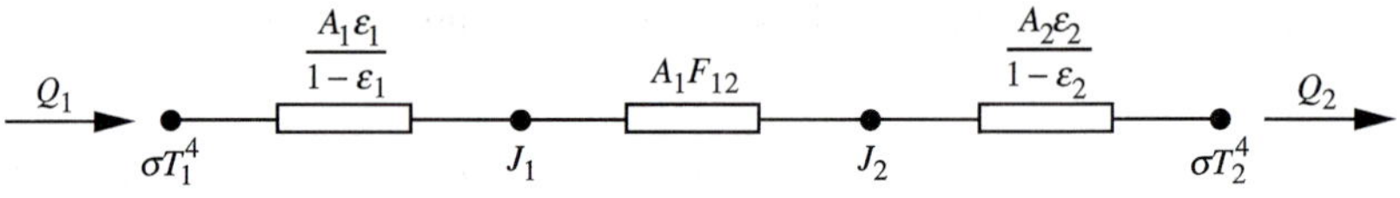

Figure 4.2   Equivalent electrical network for two grey surface enclosure.

thought of a surface conductance term which takes into account the non-ideal (non-black) of the surface.

We can now construct a network equivalent for the two grey plate problem (Figure 4.2). There will be a surface conductance term for each surface, plus the direct exchange conductance. Figure 4.2 is, in fact, applicable to any system with two grey surfaces. We are looking to determine the heat flow (current) from surface 1 to surface 2. In principle we could determine the current through any one of the three conductances; however, given that in such a problem we generally know the temperatures, the normal procedure is to determine the net conductance (C_T) between σT_1^4 and σT_2^4, i.e.

$$Q_{12} = C_T(\sigma T_1^4 - \sigma T_2^4)$$

If the temperatures are known and C_T can be found from the network, Q_{12} may be calculated.

Consider first the case of two very large grey plates. In such a system it is convenient to work with unit surface area, i.e. $A_1 = A_2 = 1$. The circuit is three conductances in series, so

$$\frac{1}{C_T} = \frac{1 - \varepsilon_1}{\varepsilon_1} + \frac{1}{F_{12}} + \frac{1 - \varepsilon_2}{\varepsilon_2}$$

Noting that $F_{12} = 1$ in this case, this can be rearranged easily to give

$$C_T = \frac{1}{\varepsilon_1^{-1} + \varepsilon_2^{-1} - 1} = \mathcal{F}_{12} \tag{4.12}$$

Thus we have reached Equations 4.10 and 4.11 with a fraction of the algebraic manipulation.

As indicated above, all two-surface problems are electrically equivalent, differing only in the magnitude of the conductances. For instance, for concentric cylinders and spheres, the net conductance is given by:

$$\frac{1}{C_T} = \frac{1 - \varepsilon_1}{A_1 \varepsilon_1} + \frac{1}{A_1 F_{12}} + \frac{1 - \varepsilon_2}{A_2 \varepsilon_2}$$

$$C_T = A_1 \left(\frac{1}{\varepsilon_1^{-1} + \frac{A_1}{A_2}(\varepsilon_2^{-1} - 1)} \right) = A_1 \mathcal{F}_{12} \tag{4.13}$$

The term in brackets is the grey radiation factor for concentric cylinders and spheres. There are three important limiting cases for this geometry.

(i)   If the separation between the surfaces is very small, then $A_1 \approx A_2$, which means that $\mathcal{F}_{12}$ reduces to that for parallel plates (Equation 4.12).

(ii)   If the outer surface is black, then $\varepsilon_2 = 1$, and $\mathcal{F}_{12} \to \varepsilon_1$.

(iii) If the separation between the surfaces is very large, then $A_1 << A_2$, and $\mathcal{F}_{12} \rightarrow \varepsilon_1$.

From cases (ii) and (iii) we conclude that if we have a small object within a very large enclosure, the walls of the enclosure appear black to the object. This is because the further away the walls, the lower the amount of radiation from the object that is reflected off the walls back to the object.

4.4 Three-surface grey enclosures—grey refractory surfaces

The three surface problem described in Section 3.2 can be extended to include grey surfaces. We will consider only a network approach for a furnace with a heat source, a heat sink, and refractory walls which are all grey. The network is shown in Figure 4.3. All pairs of surfaces are linked with conductances of the form $A_i F_{ij}$. We have surface conductance terms for the heat source and heat sink, as determined by Equation 4.4. Note that a refractory surface requires *no* surface conductance. By definition a refractory has no net flux from/to the surface, i.e. $Q_R = 0$. Thus, from Equation 4.4 we see that $J_R = \sigma T_R^4$. In other words, the radiosity of a grey refractory surface is the same as a black body at the same temperature—the emissivity of the refractory surface is immaterial.

We require the overall conductance between σT_1^4 and σT_2^4. Note that the triangle of conductances connecting the three surfaces is identical to that in the black furnace (Figure 3.3), and has a net conductance $A_1 \overline{F}_{12}$. The problem is therefore reduced to three series conductances, and we may write:

$$\frac{1}{C_T} = \frac{1 - \varepsilon_1}{A_1 \varepsilon_1} + \frac{1}{A_1 \overline{F}_{12}} + \frac{1 - \varepsilon_2}{A_2 \varepsilon_2}$$

On rearrangement we obtain:

$$C_T = A_1 \left[\frac{1}{(\overline{F}_{12})^{-1} + (\varepsilon_1^{-1} - 1) + \frac{A_1}{A_2}(\varepsilon_2^{-1} - 1)} \right] \qquad (4.14)$$

By analogy with the black furnace, we can define the term in brackets as the *total grey radiation factor*, $\overline{\mathcal{F}}_{12}$, such that $Q_{12} = C_T(\sigma T_1^4 - \sigma T_2^4) = A_1 \overline{\mathcal{F}}_{12}$. Inspection of Equation 4.14 shows that if the surfaces are black, then $\overline{\mathcal{F}}_{12} = \overline{F}_{12}$, as expected.

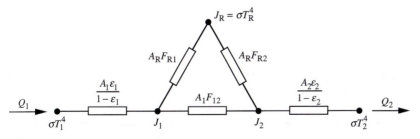

Figure 4.3 Equivalent electrical network for a three-surface grey enclosure (one heat source, one heat sink, one refractory surface).

We can find the temperature of the refractory (T_R) by carrying out appropriate current balances. For the black furnace, one balance sufficed. For the grey furnace, a current balance on the refractory surface introduces two further unknowns, J_1 and J_2:

$$A_R F_{R1}(J_1 - \sigma T_R^4) = A_R F_{R2}(\sigma T_R^4 - J_2)$$

We then require two further balances around nodes J_1 and J_2:

$$\frac{A_1 \varepsilon_1}{1 - \varepsilon_1}\left(\sigma T_1^4 - J_1\right) = A_R F_{R1}\left(J_1 - \sigma T_R^4\right) + A_1 F_{12}(J_1 - J_2)$$

$$\frac{A_2 \varepsilon_2}{1 - \varepsilon_2}\left(J_2 - \sigma T_2^4\right) = A_R F_{R2}(\sigma T_R^4 - J_2) + A_1 F_{12}(J_1 - J_2)$$

We therefore have three equations and three unknowns, which can be solved for T_R and, if required, the two radiosities. Clearly solution of this set of equations by hand is tedious. Already the limitations of the analogy are apparent: use of networks can give rates of heat transfer quite quickly, but, if temperatures or radiosities are required, we are often still required to solve a set of simultaneous equations.

4.5 Multi-surface problems

Systems involving more than three grey surfaces are rarely convenient to solve using the above methods, so we need to resort to other analytical methods. Matrix methods for sets of simultaneous equations are particularly useful. For any surface in a system, we generally know either its temperature or the heat flux to/from it. Let us recall Equations 4.2 and 4.4:

$$Q_i = \sum_j A_i F_{ij}(J_i - J_j) \tag{4.2}$$

$$Q_i = \frac{A_i \varepsilon_i}{1 - \varepsilon_i}(\sigma T_i^4 - J_i) \tag{4.4}$$

which can be combined to give for any surface i:

$$\frac{A_i \varepsilon_i}{1 - \varepsilon_i}(\sigma T_i^4 - J_i) = \sum_j A_i F_{ij}(J_i - J_j) \tag{4.15}$$

Consider each surface in turn. If its temperature is known, we can use Equation 4.15 to give an equation solely in terms of the radiosities of each surface; if the heat flux is known, we can use Equation 4.2 to give an equation which is also solely in terms of the radiosities of each surface. Thus any N-surface problem can be defined by a set of N equations with N unknown radiosities:

$$a_{11}J_1 + a_{12}J_2 + \ldots\ldots\ldots + a_{1N}J_N = C_1$$

$$a_{21}J_1 + a_{22}J_2 + \ldots\ldots\ldots + a_{2N}J_N = C_2$$

$$\vdots \qquad\qquad\qquad \vdots$$

$$a_{N1}J_1 + a_{N2}J_2 + \ldots\ldots\ldots + a_{NN}J_N = C_N$$

The coefficients a_{ij} and C_i will be functions of the known view factors and temperatures or heat fluxes. These equations can be expressed in matrix form:

$$\mathbf{AJ} = \mathbf{C}$$

$$\text{where} \quad \mathbf{A} = \begin{pmatrix} a_{11} & a_{12} & \cdots & a_{1N} \\ a_{21} & a_{22} & \cdots & a_{2N} \\ \vdots & \vdots & & \vdots \\ a_{N1} & a_{N2} & \cdots & a_{NN} \end{pmatrix} \quad \mathbf{J} = \begin{pmatrix} J_1 \\ J_2 \\ \vdots \\ J_N \end{pmatrix} \quad \mathbf{C} = \begin{pmatrix} C_1 \\ C_2 \\ \vdots \\ C_N \end{pmatrix}$$

The solution of the equations for the radiosities is obtained by inverting matrix $\mathbf{A}$, such that:

$$\mathbf{J} = \mathbf{A}^{-1}\mathbf{C}$$

$$\text{where} \quad \mathbf{A}^{-1} = \begin{pmatrix} b_{11} & b_{12} & \cdots & b_{1N} \\ b_{21} & b_{22} & \cdots & b_{2N} \\ \vdots & \vdots & & \vdots \\ b_{N1} & b_{N2} & \cdots & b_{NN} \end{pmatrix}$$

The elements of both matrices on the right-hand side of the equation are known, so each radiosity may be calculated. Once the radiosities are known, all the unknown temperatures and heat fluxes may be calculated. Matrix inversion is extremely laborious by hand, but is readily carried out on a computer.

5 Emission and absorption by gases

5.1 Introduction

So far we have assumed that the medium separating two surfaces emits no radiation itself and has no effect on the radiation passing through it, i.e. the medium is *non-participating*. The equations and methods used in the previous two chapters can usually be used to give reasonable estimates of the various unknowns. However, in many real systems the intervening medium can have a significant influence on the radiation, and some allowance must be made.

The extent to which gases emit or absorb radiation is governed by quantum physics, which is beyond the scope of this volume. There are a number of pertinent features of gas radiation:

(1) For a molecule to emit or absorb infra-red radiation, its dipole moment must change as it vibrates. So homonuclear diatomic molecules (such as N_2, O_2, H_2) do not emit or absorb, but all others do (e.g. CO, CO_2, H_2O, SO_2, NH_3, hydrocarbons). Thus systems involving air can be assumed to be non-participating, but gas radiation must be taken into account in furnaces, etc. where there are combustion products at high temperature.

(2) Radiation from solids and liquids is distributed continuously with wavelength, i.e. $\varepsilon_\lambda > 0$ for all wavelengths. For gases, radiation occurs at very specific wavelengths (or bands): emission/absorption can be very strong at some wavelengths, but can be effectively zero at others (see Figure 5.1). For example, the earth's atmosphere is transparent to visible

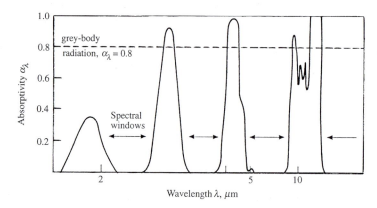

Figure 5.1 Monochromatic absorptivity for carbon dioxide.

radiation from the sun, but filters (absorbs) much of the sun's ultra-violet radiation. Thus ε_λ and α_λ are very dependent on wavelength.

(3) Radiation passing through a gas is, either transmitted, absorbed, or scattered. Scattering includes both reflection, refraction, and diffraction effects. These are especially important if the gas contains any particulate material. Scattering is discussed briefly in Chapter 9. It suffices to say at this stage that scattering can generally be neglected in many engineering systems.

(4) Gas emission and absorption is a volumetric phenomenon. The amount of absorption and emission depends on the thickness of the gas, not the contact area with surrounding surfaces.

5.2 General characteristics of gas radiation

Consider a beam of radiation of wavelength λ and initial intensity $I_{\lambda 0}$ which is incident on a layer of absorbing gas of thickness L (see Figure 5.2). The beam will be attenuated as it passes through the medium by an amount which will be proportional to the intensity of the beam, the thickness of the absorbing gas, and the concentration of the absorbing molecules (i.e. the gas pressure). Suppose that the beam intensity changes by dI_λ as the beam passes through an element of gas of thickness dx. We may write:

$$dI_\lambda = -k_\lambda p I_\lambda dx \qquad (5.1)$$

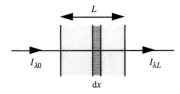

Figure 5.2 Schematic of a beam passing through a layer of absorbing gas.

where k_λ is a proportionality constant, known as either the monochromatic attenuation or extinction coefficient. This parameter includes the combined effects of absorption and scattering by the gas. (Readers should beware some ambiguities: some authors define the combination $k_\lambda p$ as the extinction coefficient; there is also a tendency to regard the extinction coefficient and absorption coefficient to be synonymous—this is true only if scattering is negligible.) k_λ is a property of the gas and measures the absorbing and scattering power of the gas. It is wavelength-dependent, and also varies with the temperature of the gas. In a mixture of absorbing and non-absorbing gases (e.g. CO_2 in air) p can be taken to be the partial pressure of the absorbing species.

If the gas is isothermal, then k_λ can be assumed to be independent of x, and the integration of Equation 5.1 over the length of the gas layer gives:

$$\int_{I_{\lambda 0}}^{I_{\lambda L}} \frac{dI_\lambda}{I_\lambda} = -\int_0^L k_\lambda p \, dx$$

$$\Rightarrow \quad \frac{I_{\lambda L}}{I_{\lambda 0}} = e^{-k_\lambda p L}$$

This expression is known as Beer's Law, and predicts exponential decay of the beam intensity as it passes through the gas. The ratio $(I_{\lambda L}/I_{\lambda 0})$ represents

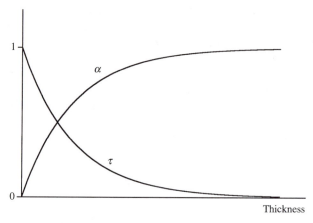

Figure 5.3 Absorptivity and transmissivity of an absorbing medium as a function of thickness.

the fraction of incident radiation of wavelength λ which is transmitted through the gas, i.e. the monochromatic transmissivity, τ_λ:

$$\frac{I_{\lambda L}}{I_{\lambda 0}} = \tau_\lambda = e^{-k_\lambda pL} \tag{5.2}$$

Assuming negligible scattering, the monochromatic absorptivity is given by:

$$\alpha_\lambda = 1 - \tau_\lambda = 1 - e^{-k_\lambda pL} \tag{5.3}$$

Applying Kirchhoff's Law (Equation 1.18) gives the monochromatic emissivity:

$$\varepsilon_\lambda = \alpha_\lambda = 1 - e^{-k_\lambda pL} \tag{5.4}$$

For small values of $k_\lambda pL$, $\alpha_\lambda \approx k_\lambda pL$, i.e. the absorptivity varies linearly with L. Figure 5.3 plots schematically α_λ and τ_λ as a function of the thickness of the absorbing layer. Media where α_λ, and hence ε_λ, are both close to zero, i.e. the material is almost transparent, are generally said to be *optically thin*; media where $\tau_\lambda \to 0$ are said to be *optically thick*. However, caution needs to be exercised with these terms, because the band nature of emission/absorption from gases means that a material will be optically thin at some wavelengths, but optically thick at others.

5.3 Mean beam length

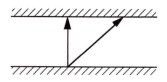

Figure 5.4 Schematic showing the need for a mean beam length.

The algebra in the previous section was for a one-dimensional system of thickness L. Most real systems are two or three dimensional, which poses a problem of how to define L. For example, Figure 5.4 shows two parallel plates, with two beams of radiation originating from the same point on one plate and travelling to the opposite plate. Clearly the thickness of gas encountered by each beam is different, as, indeed, it will be for every other beam travelling between the two plates. So, how should we define L? We need to introduce the idea of the *mean beam length*, which represents an average thickness of gas encountered by all beams of radiation between the two plates.

Consider an enclosure of volume V and surface area A with black walls (Figure 5.5). Let dA be an area element, from which a beam of radiation leaves at an angle θ to the normal at dA. At a distance x from dA let the beam meet a volume element of gas dV, which has length dx and cross-section area dS normal to the beam.

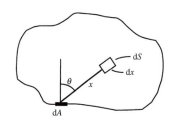

Figure 5.5 Radiative transfer from a surface area element to a gas volume element in an enclosure.

Using the Cosine Law (Equation 2.5), the radiation of wavelength λ emitted by dA towards dS is:

$$\frac{W_{b\lambda} d\lambda}{\pi} \cdot \frac{dS \, dA \cos \theta}{x^2}$$

The beam will be attenuated as it travels through the gas. Thus, from Equation 5.2, the radiation entering the gas element will be:

$$\frac{W_{b\lambda} d\lambda}{\pi} \cdot \frac{dS \, dA \cos \theta}{x^2} \cdot e^{-k_\lambda px}$$

Using Equation 5.1 the radiation absorbed by the gas element is:

$$k_\lambda p \frac{W_{b\lambda} d\lambda}{\pi} \cdot \frac{dS \, dA \cos \theta}{x^2} \cdot e^{-k_\lambda px} \, dx = k_\lambda p \frac{W_{b\lambda} d\lambda}{\pi} \cdot \frac{dA \cos \theta}{x^2} \cdot e^{-k_\lambda px} \, dV$$

We can integrate over the whole of the surface of the enclosure to give the absorption by the gas element of all incident radiation of wavelength λ:

$$Q_\lambda = \frac{k_\lambda p W_{b\lambda} \, d\lambda \, dV}{\pi} \int_A \frac{\cos \theta}{x^2} \cdot e^{-k_\lambda px} dA \qquad (5.5)$$

This is generally very difficult to integrate. In optically thin media, $k_\lambda px$ is small, such that $e^{-k_\lambda px} \approx 1$. Equation 5.5 then becomes:

$$Q_\lambda = \frac{k_\lambda p \, W_{b\lambda} \, d\lambda \, dV}{\pi} \int_A \frac{\cos \theta}{x^2} dA \qquad (5.6)$$

Now, $(\cos \theta \, dA/x^2)$ is the solid angle $d\Omega$ subtended by dA at dV, so we can rewrite Equation 5.6:

$$Q_\lambda = \frac{k_\lambda p W_{b\lambda} \, d\lambda \, dV}{\pi} \int_A d\Omega = 4 k_\lambda p W_{b\lambda} d\lambda \, dV$$

given that the total solid angle is 4π for an enclosure. This can be integrated over the whole volume to give:

$$\text{Absorption} = 4 k_\lambda p W_{b\lambda} \, V \, d\lambda$$

$$\text{Absorption per unit area of emitting surface} = \frac{4 k_\lambda p W_{b\lambda} \, V \, d\lambda}{A} \qquad (5.7)$$

This gives a general expression for any enclosure. Let us compare this with a known geometry, viz. a hemisphere of radius R, with radiation of wavelength

Table 5.1   Mean Beam Lengths for gas radiation in a number of geometries

Geometry	$4V/A$	$3.6V/A$
Sphere (radius r) radiating to surface	$\dfrac{4 \times \frac{4}{3}\pi r^3}{4\pi r^2} = \frac{4}{3}r$	$1.2r$
Cube (side length L) radiating to one face	$\dfrac{4 \times L^3}{L^2} = 4L$	$3.6L$
Cube (side length L) radiating to all faces	$\dfrac{4 \times L^3}{6L^2} = \frac{2}{3}L$	$0.6L$
Cylinder (radius r, length L) radiating to curved surface	$\dfrac{4 \times \pi r^2 L}{2\pi rL} = 2r$	$1.8r$
Cylinder (radius r, length L), radiating to curved surface and ends	$\dfrac{4 \times \pi r^2 L}{2\pi rL + 2\pi r^2} = \dfrac{2rL}{L+r}$	$\dfrac{1.8rL}{L+r}$
Space between two parallel plates (each area A, distance L apart) to both plates	$\dfrac{4 \times AL}{2A} = 2L$	$1.8L$

λ from a surface element at the centre of the base to the whole of the curved surface. In this case all beams are of length R. From Equation 5.3:

$$\text{Absorption per unit area of emitting surface} = W_{b\lambda}\,\mathrm{d}\lambda(1 - e^{-k_\lambda pR})$$

$$\approx W_{b\lambda}k_\lambda pR\,\mathrm{d}\lambda \quad \text{for small } k_\lambda pR \tag{5.8}$$

A comparison of Equations 5.7 and 5.8 shows that:

$$\text{Beam Length }(R) = \frac{4V}{A} = \frac{4 \times \text{Volume of gas}}{\text{Area of contact between gas and surface}}$$

This has been derived for an optically thin gas. More detailed analysis for gases with finite values of $k_\lambda pR$ shows that this expression overestimates the actual beam length by around 10%, because a significant proportion of the radiation emitted by the gas is absorbed elsewhere in the gas before it reaches the surface. In practical systems it is therefore usual to use

$$\text{Beam Length} \quad \approx \quad \frac{3.6V}{A}$$

Although derived for a hemisphere, it has become customary for beam lengths for all shapes of enclosure to be defined in similar terms. Thus we have:

$$\text{Mean Beam Length }(L_\mathrm{m}) = \frac{3.6 \times \text{Volume of gas}}{\text{Area of contact between gas and surface}} \tag{5.9}$$

This can be evaluated quite easily for many geometries, and tabulations are given in many sources. A small selection is given in Table 5.1.

5.4   Total emissivity for a gas

The concept of total emissivity introduced in Chapter 1 in the context of surfaces can be applied equally well to gases. Using the previous definitions, we can define the total emissivity of a gas ε_g:

$$\varepsilon_g = \frac{W}{W_b} = \frac{1}{\sigma T_g^4} \int_0^\infty \varepsilon_\lambda W_{b\lambda} \, d\lambda$$

where T_g is the absolute temperature of the gas. Substituting Equation 5.4 for ε_λ and Planck's Law (Equation 1.13) for $W_{b\lambda}$, we obtain:

$$\varepsilon_g = \frac{1}{\sigma T_g^4} \int_0^\infty \left[1 - e^{-k_\lambda p L_m}\right] \left[\frac{2\pi h c^2}{\lambda^5}\left(\frac{1}{\exp\left(\frac{hc}{\lambda k T_g}\right) - 1}\right)\right] d\lambda \qquad (5.10)$$

We can also define the total gas emissivity in terms of an overall absorption coefficient (k):

$$\varepsilon_g = 1 - e^{-k p L_m} \qquad (5.11)$$

It can be seen that ε_g is a function of the gas temperature and the product pL_m. The integral in Equation 5.10 can, in principle, be evaluated, provided that the wavelength dependence of the absorption coefficient is known. This has been done by a number of workers, using experimental spectral data. For engineering calculations, it is generally most convenient if values of ε_g can be presented in either tabular or graphical form. Hottel's pioneering experimental work included the production of emissivity charts for a number of gases of practical importance (see, for example, McAdams). More recent work combining theory and experiment has produced more refined data over a much wider range of conditions (see review by Modest). However, Hottel's charts still find favour among many engineers. Agreement between Hottel's charts and more recent data is generally good, except that the charts should not be extrapolated outside the limits of Hottel's data.

Figures 5.6 and 5.7 are emissivity charts for carbon dioxide and water vapour at 1 bar total pressure as a function of temperature for various values of pL_m. The total emissivity is simply read directly from the chart, given knowledge of the gas temperature, the partial pressure of the absorbing gas, and the mean beam length for the system. These charts can be used for pure CO_2 or pure water vapour, or mixtures of either gas in a non-participating gas, such as air. Mixtures of two participating gases (e.g. combustion products containing CO_2 and H_2O) will be considered in Section 5.6.

For total pressures other than 1 bar, Hottel produced pressure correction charts. These give a correction factor (C) which must be multiplied with the value of emissivity for 1 bar pressure, i.e. $\varepsilon(p_{total}) = C \times \varepsilon(1 \text{ bar})$. More modern data are available, including empirical correlation equations, but charts are still considered to be more convenient for order-of-magnitude engineering calculations. Again, they should not be used outside the limits of Hottel's original data. Figure 5.8 shows a pressure correction chart for carbon dioxide, and clearly shows that emissivity generally increases as total pressure increases.

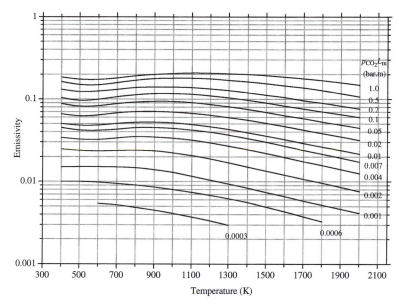

Figure 5.6. Emissivity chart for carbon dioxide at 1 bar total pressure as a function of temperature and $P_{CO_2}L_m$ (redrawn from Siegel and Howell).

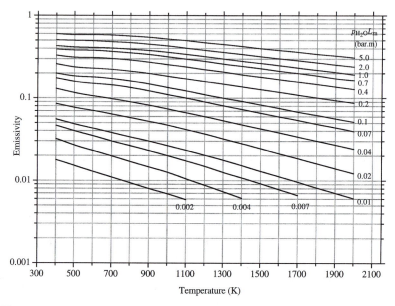

Figure 5.7 Emissivity chart for water vapour at 1 bar total pressure as a function of temperature and $P_{H_2O}L_m$ (redrawn from Siegel and Howell).

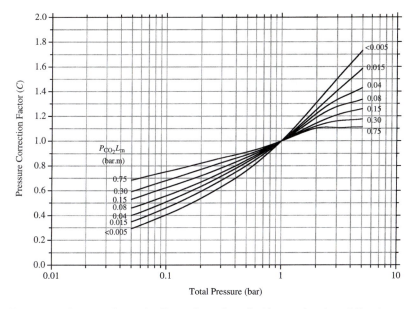

Figure 5.8   Pressure Correction Factor for carbon dioxide as a function of $P_{CO_2}L_m$.

5.5   Total absorptivity

The defining equations for absorption can also be applied to gases. Thus if a gas receives radiation from a surface whose temperature is T_s, then, from Equation 1.17, the total absorptivity α_g for the gas is given by:

$$\alpha_g = \frac{\int_0^\infty \alpha_\lambda G_\lambda d\lambda}{\int_0^\infty G_\lambda \, d\lambda} = \frac{1}{\sigma T_s^4}\int_0^\infty \alpha_\lambda G_\lambda \, d\lambda$$

$$= \frac{1}{\sigma T_s^4}\int_0^\infty \left[1 - e^{-k_\lambda p L_m}\right]\left[\frac{2\pi hc^2}{\lambda^5}\left(\frac{1}{\exp\left(\frac{hc}{\lambda k T_s}\right) - 1}\right)\right]d\lambda \qquad (5.12)$$

The right-hand side of Equation 5.12 shows that total absorptivity is a function of pL_m, T_g, and T_s. The dependence on the temperature of the surrounding surfaces creates a major complication if this integral is to be evaluated. There is, however, an empirical short-cut, originally developed by Hottel, which uses the total emissivity charts (Figure 5.6 and 5.7) to determine the total absorptivity. A value for α_g can be found using the following procedure:

(i)   use T_s not T_g on the horizontal axis;
(ii)   use $pL_m \times (T_s/T_g)$ instead of pL_m;
(iii)   read off a value from the chart;
(iv)   multiply by $(T_g/T_s)^{0.5}$.

Hottel's pioneering work recommended (on purely empirical grounds) using exponents of 0.65 for CO_2 and 0.45 for H_2O in step (iv). Although these values are still in common use, a single value of 0.5 can be justified theoretically.

Example 5.1.
CO_2 at 1 bar and 1500 K is in a vessel whose walls are at 1000 K. If pL_m is 0.02 bar m, determine ε_g and α_g.

Solution.
From Figure 5.6, ε_g ($T_g = 1500$ K, $pL_m = 0.02$ bar m) = **0.051**
$$\alpha_g = \varepsilon_g \ (1000 \text{ K}, \ 0.02 \times (1000/1500) \text{ bar m}) \times (1500/1000)^{0.5}$$
$$= \varepsilon_g \ (1000 \text{ K}, \ 0.0133 \text{ bar m}) \times (1500/1000)^{0.5}$$
$$= 0.056 \times (1500/1000)^{0.5}$$
$$= \mathbf{0.069}$$

Note that the total emissivity and absorptivity have quite different values. Although this is a simple example, it illustrates a common complication with gas radiation, viz. $\varepsilon_g \neq \alpha_g$. This means that gases cannot normally be assumed to be grey. We shall see later in Section 5.8 how using an assumption of grey gases results in easier calculations, but can give rise to errors which are unacceptable in all but rough calculation.

5.6   Spectral overlap

In systems where more than two absorbing/emitting species are present, there will be a contribution from each component. A straight addition of the individual contributions will not generally suffice, because of overlap of the spectral bands of the various species, i.e. two or more different species having radiation bands at the same wavelength. The effect of overlap is increased self-absorption, which lowers the emissivity of the gas mixture, compared with the sum of the individual contributions.

The usual approach is to find the individual contributions from charts, etc., and then to apply an overlap correction. So, for a mixture of two gases we can write:

$$\varepsilon_{\text{mixture}} = \varepsilon_{g1} + \varepsilon_{g2} - \Delta\varepsilon \tag{5.13}$$

where ε_{g1} and ε_{g2} are the contributions from each gas (each being determined as if the other gas were absent) and $\Delta\varepsilon$ is the spectral overlap correction. Hottel originally produced empirical charts of $\Delta\varepsilon$ for various CO_2/H_2O mixtures as a function of temperature (see Figure 5.9). A considerable degree of interpolation is required with these charts. However, the uncertainty which this introduces is negligible, as $\Delta\varepsilon$ is normally small compared with the sum of ε_{g1} and ε_{g2}. Indeed, for CO_2/H_2O mixtures, neglecting spectral overlap completely introduces an error of only 10 to 15% at most in practical systems.

We can use an analogous procedure to determine spectral overlap corrections for the total absorptivity of mixtures. Thus for a mixture of two gases we have:

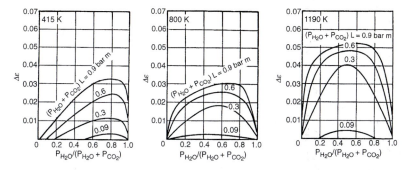

Figure 5.9 Spectral overlap correction for gas mixtures containing H_2O and CO_2.

$$\alpha_{\text{mixture}} = \alpha_{g1} + \alpha_{g2} - \Delta\alpha$$

where α_{g1} and α_{g2} are the contributions from each gas and $\Delta\alpha$ is the spectral overlap correction. $\Delta\alpha$ is determined from the $\Delta\varepsilon$ charts, using the temperature of the radiation source, instead of the gas temperature.

It is obvious that determination of gas emissivities and absorptivities is somewhat approximate. Emissivities are generally difficult to measure, so all published charts include a degree of uncertainty. Recent workers have attempted to derive correlations for the experimental data, particularly as an aid for computer-based calculations. Such correlations normally involve too may parameters to help in simple hand-calculations, and are not therefore included here.

5.7 Network analogy for gases

We have seen how radiation heat transfer problems are solved much more easily if the network approach is used, especially if the grey body approximation is valid. The network approach can also be used for participating media, provided that we can take account of the effect of the medium on transfer between surfaces. We need to consider the effect of absorption on the direct transfer of radiation between the two surfaces, and the emission of radiation by the gas to each surface.

Consider two surfaces separated by an absorbing/emitting gas. Then:

Transmission from surface 1 to surface 2 $= J_1A_1F_{12}\tau_{g1} = J_1A_1F_{12}(1 - \alpha_{g1})$

Transmission from surface 2 to surface 1 $= J_2A_2F_{21}\tau_{g2} = J_2A_2F_{21}(1 - \alpha_{g2})$

where τ_{gi} is the transmissivity of the gas to the radiation emitted by surface i. Note that because the spectral distribution of the radiation from each surface is different, the proportion of the radiation originating from each surface which is transmitted/absorbed will also differ. The net transmission Q_{12} from surface 1 to surface 2 will be, after using reciprocity:

$$Q_{12} = A_1F_{12}\big[J_1(1 - \alpha_{g1}) - J_2(1 - \alpha_{g2})\big]$$

Let us assume that the gas is grey, i.e. whatever the radiation source, the absorptivity equals the emissivity then $\alpha_{g1} = \varepsilon_g$ and $\alpha_{g2} = \varepsilon_g$, which means that:

$$Q_{12} = A_1 F_{12}(1 - \varepsilon_g)(J_1 - J_2) \tag{5.14}$$

This is now in a form which is suitable for use in the network analogy. On comparison with Equation 4.1 for transfer between non-black surfaces, it is clear that the effect of the gas is to modify the conductance between nodes J_1 and J_2 by a factor $(1 - \varepsilon_g)$.

We must now consider radiation to/from the gas. For interchange between the gas and surface 1, we have:

$$\text{Emission from gas to surface } 1 = A_1 \varepsilon_g \sigma T_g^4$$

$$\text{Radiation from surface 1 absorbed by gas} = A_1 J_1 \alpha_{g1}$$

As the gas is in intimate contact with the surface, the view factor between the surface and the gas can be taken to be 1. The net radiative transfer from the gas to surface 1 is therefore:

$$Q_{g1} = A_1(\varepsilon_g \sigma T_g^4 - \alpha_{g1} J_1^4)$$

If the gas can be assumed to be grey, then $\alpha_{g1} = \varepsilon_g$ and the equation becomes

$$Q_{g1} = A_1 \varepsilon_g(\sigma T_g^4 - J_1^4) \tag{5.15}$$

Once again an assumption of a grey gas has given a form which is suitable for the network analogy. Transfer between a surface and a gas can be represented by a conductance $A_1 \varepsilon_g$ between a σT_g^4 node for the gas and a radiosity node for the surface. Note that the gas is similar to a refractory in its use of the emissive power as the voltage node.

5.8 Worked example

In order to demonstrate the use of gas emissivity data in engineering calculations, this section considers a simple geometry (two large parallel plates) using a variety of different assumptions and conditions. Figure 5.10 shows the system: the two plates are 2 m apart and at temperatures of 1800 K and 1000 K. The gas between the plates is at a constant, uniform temperature and is stagnant, such that convective effects can be ignored. Thermal

Figure 5.10 Geometry for worked example for gas radiation.

conduction through the gas will be assumed to be negligible. Four different conditions will be considered:

(i)   black surfaces, non-participating gas;
(ii)   black surfaces, participating gas (20 vol% CO_2 in air at 1 bar);
(iii) black surfaces, grey gas;
(iv)   grey surfaces, grey gas.

In each case we will determine the rate of heat transfer from the hotter to the cooler plate, and the temperature of the gas.

Case I—Black surfaces, non-participating gas
If the gas is non-participating, then it has no rôle in the problem, which is reduced to a simple two-surface problem of the type discussed in Chapter 3. As the plates are very large, it is convenient to determine the heat flux, which will be given by:

$$\text{Flux} = F_{12}\sigma(T_1^4 - T_2^4) = 1 \times 5.67 \times 10^{-8} \times (1800^4 - 1000^4)$$

$$= \textbf{539 kW/m}^2$$

Since the gas plays no rôle, its temperature is irrelevant to the analysis, and the calculated heat flux holds for all gas temperatures. The corollary to this is that a value for the gas temperature is indeterminate. The gas temperature is relevant (and hence calculable) only if the assumption that convection is negligible breaks down, or if the gas absorbs/emits radiation.

Case II—Black surfaces, participating gas
The problem is most easily solved by considering the net radiant heat flux (Q_1) leaving the hot plate, which, because of the geometry of the problem, is equal to the net radiant heat flux arriving at the cooler plate. A balance on surface 1 gives:

$$Q_1 = \text{Emission from plate } 1 - (\text{Emission from gas to plate } 1 + \text{Emission from plate 2 which reaches plate 1})$$

$$Q_1 = A_1\sigma T_1^4 - A_1\varepsilon_g\sigma T_g^4 - A_2 F_{21}\sigma T_2^4 \tau_{g2}$$

Using reciprocity, and noting that $F_{12} = 1$, the heat flux becomes

$$\text{Flux} = \sigma T_1^4 - \varepsilon_g\sigma T_g^4 - \sigma T_2^4(1 - \alpha_{g2}) \tag{5.16}$$

This gives us two unknowns—the flux and the gas temperature T_g. A second equation can be obtained from a balance on the gas. The gas is neither a heat source nor a heat sink, so can be considered to be essentially adiabatic, in which case:

Radiation absorbed from surface 1 + Radiation absorbed from surface 2
= Radiation emitted by gas to both surfaces

$$A_1\sigma T_1^4\alpha_{g1} + A_2\sigma T_2^4\alpha_{g2} = (A_1 + A_2)\sigma T_g^4\varepsilon_g$$

Rearrangement, noting that $A_1 = A_2$ for parallel plates, gives:

$$T_g^4 = \left(\frac{\alpha_{g1}T_1^4 + \alpha_{g2}T_2^4}{2\varepsilon_g}\right) \tag{5.17}$$

So, the problem is solved if we determine the emissivity and absorptivities, calculate the gas temperature using Equation 5.17, and then the flux using Equation 5.16.

First, we need to find the mean beam length for the system. For unit cross-section, the volume is 2 m^3, and the surface area to which the gas radiates is 2 m^2. Therefore using Equation 5.9 gives

$$L_m = \frac{3.6V}{A} = 3.6 \text{ m}$$

$$\Rightarrow \quad p_{CO_2}L_m = 0.2 \times 3.6 = 0.72 \text{ bar m}$$

Use of the emissivity chart requires knowledge of the gas temperature, which is currently unknown. We must therefore make an initial estimate of the gas temperature in order to determine emissivities and absorptivities. We can then calculate the gas temperature using Equation 5.17; if our initial estimate of T_g is wrong, we can iterate to give a better value. In this example, an initial estimate halfway between the temperature of the two surfaces would seem to be a sensible starting point, so let us assume a gas temperature of 1400 K. (Note that Figure 5.6 shows that ε_g for CO_2 is not a strong function of temperature in the range 1300 to 1600 K, so the problem will not be greatly affected by any error in this estimate.) From Figure 5.6:

$$\varepsilon_g(1500 \text{ K}, 0.72 \text{ bar m}) = 0.18$$

We now use the procedure of Section 5.5 to determine absorptivities. For α_{g1}:

$$T_1 = 1800 \text{ K} \quad \text{and} \quad pL_m\left(\frac{T_1}{T_g}\right) = 0.72 \times \frac{1800}{1400} = 0.93 \text{ bar m}$$

$$\alpha_{g1} = \varepsilon_g(1800 \text{ K}, 0.93 \text{ bar m}) \times \left(\frac{1400}{1800}\right)^{0.5}$$

$$= 0.16 \times 0.88 = 0.14$$

Similarly for α_{g2}:

$$T_2 = 1000 \text{ K} \quad \text{and} \quad pL_m\left(\frac{T_2}{T_g}\right) = 0.72 \times \frac{1000}{1400} = 0.51 \text{ bar m}$$

$$\alpha_{g2} = \varepsilon_g(1000 \text{ K}, 0.51 \text{ bar m}) \times \left(\frac{1400}{1000}\right)^{0.5}$$

$$= 0.18 \times 1.18 = 0.21$$

Substitution of emissivity and absorptivities into Equation 5.17 gives:

$$T_g = \left(\frac{0.14 \times 1800^4 + 0.21 \times 1000^4}{2 \times 0.18}\right)^{\frac{1}{4}} = \textbf{1470 K}$$

Given the two figure accuracy of the charts and the weak temperature dependence of ε_g, this is probably close enough to our initial estimate not to require any iteration. Substitution into Equation 5.16 gives:

$$\text{Heat Flux} = 5.67 \times 10^{-8}\left(1800^4 - 0.18 \times 1470^4 - 1000^4(1 - 0.21)\right)$$
$$= \mathbf{503\ kW/m^2}$$

Compared with Case I, we see that the effect of the gas is to reduce the net heat flux by almost 7%. Note that surface 1 is still radiating at the same rate as before—it is still a black surface at 1800 K. Energy is not being lost, but is being absorbed by the gas and re-radiated back to surface 1, thereby reducing the next flux from surface 1 to surface 2.

Case III—Black surfaces, grey gas

For rough calculations it can be time-consuming to carry out the full analysis of case II. We can get a reasonable estimate of heat flux if we assume that the gas is grey, even though it is clear that α_{g1}, α_{g2}, and ε_g are not very similar. The heat balances in Equations 5.16 and 5.17 remain the same as before; the only difference in the solution is that we assume that $\alpha_{g1} = \alpha_{g2} = \varepsilon_g = 0.18$. Therefore:

$$T_g = \left(\frac{0.18 \times 1800^4 + 0.18 \times 1000^4}{2 \times 0.18}\right)^{\frac{1}{4}} = \mathbf{1548\ K}$$

$$\text{Heat Flux} = 5.67 \times 10^{-8}\left(1800^4 - 0.18 \times 1548^4 - 1000^4(1 - 0.18)\right)$$
$$= \mathbf{490\ kW/m^2}$$

Clearly the assumption that the gas is grey introduces a sizeable overestimate of the gas temperature, and a slight underestimate of the heat flux.

Case IV—Grey surfaces, grey gas

We have seen in Chapters 3 and 4 how use of the electrical network analogy can increase the speed of solution of radiation heat transfer problems. The case of two grey surfaces and a grey gas is most easily addressed using such a network. Figure 5.11 shows the network appropriate for our geometry, using the conductances derived in Section 5.7. We shall assume in this problem that $\varepsilon_1 = 0.75$ and $\varepsilon_2 = 0.8$.

We note, as before, that $A_1 = A_2$ and $F_{12} = 1$ for large parallel plates, and that $\varepsilon_g = 0.18$. We require the overall conductance between σT_1^4 and σT_2^4. The

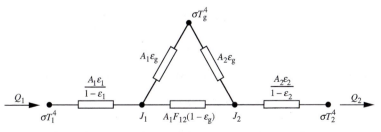

Figure 5.11 Equivalent electrical network for two grey surfaces and a grey gas.

triangle of conductances between J_1 and J_2 has net conductance

$$A_1 F_{12}(1 - \varepsilon_g) + \cfrac{1}{\left(\cfrac{1}{A_1 \varepsilon_g} + \cfrac{1}{A_2 \varepsilon_g}\right)} = A_1\left(1 - \frac{\varepsilon_g}{2}\right)$$

and the overall conductance C_T is given by

$$\frac{1}{C_T} = \frac{1 - \varepsilon_1}{A_1 \varepsilon_1} + \cfrac{1}{A_1\left(1 - \cfrac{\varepsilon_g}{2}\right)} + \frac{1 - \varepsilon_2}{A_2 \varepsilon_2}$$

which, on substitution of the emissivity data, leads to $C_T = 0.594 A_1$. The rate of heat transfer is given by

$$Q_{12} = C_T(\sigma T_1^4 - \sigma T_2^4)$$

$\Rightarrow$ Heat Flux $= 0.594 \times 5.67 \times 10^{-8} \times (1800^4 - 1000^4) = $ **320 kW/m²**

Equation 4.4 can be used to give $J_1 = 489$ kW/m² and $J_2 = 137$ kW/m². A current balance around the gas node gives

$$A_1 \varepsilon_g\left(J_1 - \sigma T_g^4\right) = A_2 \varepsilon_g\left(\sigma T_g^4 - J_2\right)$$

$$T_g = \textbf{1532 K}$$

The effect of having grey rather than black surfaces is a large reduction in the net rate of heat transfer, but, in this example, there is relatively little effect on the temperature of the gases.

6 Radiative heat transfer in furnaces

6.1 Introduction

We turn now to one important application of radiation heat transfer, namely the high temperature furnace, where radiation is generally the dominant mechanism of heat transfer. The object of any furnace is to heat some material (generally known as the *stock*) using a high temperature heat source, such as combustion gases or an electrically heated panel. A wide variety of different geometries can be used, depending on the application. Correct design and sizing of a furnace depends on accurate modelling of radiation heat transfer within the furnace.

This chapter considers two aspects of radiative heat transfer calculations which are particularly pertinent to furnaces, although the principles described are equally applicable to other systems. First we consider the question of time dependence: furnace designers are frequently concerned with the heating time required for a given material to reach a target temperature. Second we address the non-isothermal nature of most furnaces: in general we cannot assume that the walls and gases are at a constant temperature, and some allowance must be made for this.

6.2 Variation of temperature with time

All analyses so far have assumed steady-state conditions, where surface and gas temperatures do not vary with time. Many applications involve placing material in a furnace and waiting until it has reached some pre-determined temperature. A furnace will normally be designed such that the approach of the stock to the target temperature will not normally be asymptotic. This requires the heat source to be at a temperature considerably higher than the target temperature. If the approach to the target temperature is asymptotic, then the length of time for which the material is at high temperature may be long enough for appreciable oxidation to occur, which may lead to impaired product quality. It is therefore important that designers and operators are able to predict temperature–time histories for furnaces. We consider only a very simple example here, but the principles can be applied to more complex models.

Consider a rectangular furnace with grey surfaces, consisting of a heater panel (surface 1) at constant temperature T_1, the stock (surface 2) at variable temperature T placed on the floor of the furnace, and all remaining surfaces refractory. If the heater panel is electrically heated, then we can assume that the furnace gases are primarily air, so gas radiation can be ignored. (In a fossil fuel-fired furnace, the hot combustion gases would be the heat source, and it can normally be assumed that all surfaces, other than the stock, are refractory.)

Let the stock have mass m and specific heat capacity C. We shall also assume that the stock has a sufficiently high thermal conductivity that the stock is at a uniform temperature and there are no temperature gradients within it. If heat transfer is by radiation alone, then the instantaneous net rate of heat transfer from the heat source to the stock is:

$$Q_{12} = A_1 \overline{\mathcal{F}}_{12}(\sigma T_1^4 - \sigma T^4) \tag{6.1}$$

where $\overline{\mathcal{F}}_{12}$ is the total grey radiation factor for transfer from the heat source to the stock. $\overline{\mathcal{F}}_{12}$ includes all the information about the geometry (via the view factors) and emissivities of the surfaces. Equation 6.1 can be applied to any multi-surface furnace, not just the three-surface example given here. The limiting factor is the ease of calculation of $\overline{\mathcal{F}}_{12}$: as we have seen in Chapter 4, the more complex the geometry, the more difficult $\overline{\mathcal{F}}_{12}$ is to calculate.

Consider also a time interval dt in which heat dH is transferred resulting in a temperature rise dT. We can write:

$$dH = mCdT$$

$$Q_{12} = \frac{dH}{dt} = mC\frac{dT}{dt} \tag{6.2}$$

Combining Equations 6.1 and 6.2 gives:

$$\frac{dT}{dt} = \frac{\sigma A_1 \overline{\mathcal{F}}_{12}}{mC}(T_1^4 - T^4)$$

Let us assume that the stock is heated from an initial temperature T_0 at $t = 0$ to a general temperature T at time t. The differential equation may be solved by separating the variables and integrating, assuming that the specific heat capacity of the stock and all emissivities (hence $\overline{\mathcal{F}}_{12}$) do not vary with temperature:

$$\int_0^t dt = \frac{mC}{\sigma A_1 \overline{\mathcal{F}}_{12}} \int_{T_0}^T \frac{dT}{(T_1^4 - T^4)}$$

The right-hand side is a standard integral, from which we obtain the solution:

$$t = \frac{mC}{\sigma A_1 \overline{\mathcal{F}}_{12}} \cdot \frac{1}{4T_1^3}\left[\ln\left|\frac{T_1 + T}{T_1 - T}\right| + 2\tan^{-1}\left(\frac{T}{T_1}\right)\right]_{T=T_0}^{T=T} \tag{6.3}$$

Equation 6.3 gives the temperature-time history for the material being heated. Figure 6.1 plots the equation for two values of T_1 (1200 and 1500 K), with an initial temperature (T_0) of 298 K. The asymptotic nature of the heating curve is evident: the curve begins to flatten appreciably at around 300 K below the temperature of the heat source. The most effective heating (in terms of high rate of rise of temperature) is therefore obtained if the temperature of the source is significantly above the target temperature of the stock. If the stock is

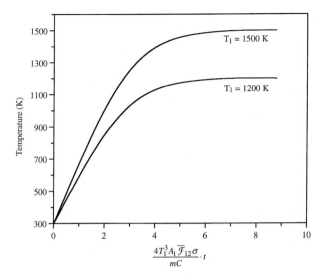

Figure 6.1 Temperature–time plot for a furnace with heat sources of two different temperatures.

likely to be affected by excessive residence time at high temperature, the designer should ensure that a sufficiently high temperature source is used so that heating occurs on the linear part of the heating curve. A thorough review of furnace design is given by Rhine and Tucker.

6.3 Non-Isothermal systems

All analyses so far have assumed that a participating gas is isothermal and that all surfaces are at a uniform temperature. Most furnaces are far from this idealized view. For example, if a box furnace of the type introduced in Chapter 3 had walls of uniform temperature, then there would be a discontinuity in the temperature where one surface met another; similarly, when radiating gases were discussed in Chapter 5, the gas and all surfaces were assumed to be uniform, which results in a temperature discontinuity at the interface between the gas and any surface. In real systems there are no discontinuities, but a continuous change in temperature at the interface between two surfaces or a surface and a gas.

If we require accurate calculation, then this issue needs to be addressed, and a number of methods have been developed which attempt to reconcile the non-isothermal nature of real surfaces and gases. (A thorough account of many of the available methods is given by Modest.) Four generic types of heat transfer analysis are in use. The *spherical harmonics method* and the *discrete ordinates method* both attempt to transform the radiative transfer equations into a set of simultaneous partial differential equations, which can then be solved usually by numerical procedures; the *zone method* subdivides a non-isothermal system into a series of isothermal zones, which can then be analysed using the methods already described; the *Monte Carlo method* extends the statistical techniques introduced for view factor determination in Section 2.7 to a complete analysis of tracing the history of randomly generated beams of radiation as they travel through an enclosure. A number of hybrid schemes, involving a mixture of two or more these methods, have been proposed within the last decade or so, e.g. the discrete transfer method of Lockwood and Shah. We shall concentrate on what was historically the first approach to be used successfully in furnaces, and is arguably the most suited to hand-calculation. This is the zonal method, which was originally developed by Hottel.

As indicated above, the basis of the method is to divide the non-isothermal system into a sufficiently large number of zones, such that each zone can be considered to be essentially isothermal. Surfaces can be sub-divided into a number of surface zones; gas volumes can be sub-divided into a number of volume zones. The problem is thereby reduced to a N-surface problem of the type discussed in earlier chapters: heat balances can be set up resulting in a set of N simultaneous equations which can be solved using matrix algebra. Clearly, the greater the number of sub-divisions, the greater the number of zones and the more accurate the results (i.e. the better the spatial resolution). The procedure can be summarised in the following way:

(i)   divide system into a suitable number of zones;
(ii)   assume wall and gas temperatures are uniform in each zone;
(iii) neglect conduction and convection effects;
(iv)   perform a heat balance on each zone, and solve for the unknowns.

Neglecting conduction and convection is a fair assumption in most high temperature furnaces, unless gas velocities over the surfaces are high. If required, it is relatively straightforward to allow for convection and conduction when constructing the heat balances. Simultaneous conduction and convection are discussed in detail in Chapter 7.

Very often the zonal method can result in a sequential, rather than simultaneous set of N equations, which makes hand-calculation very much easier. This is illustrated in the following example, where we wish to determine the temperature distribution in a long furnace.

Example 6.1.
A furnace 5 m long with 1 m × 1 m cross-section is fed with a fuel/air mixture at 2 kg/s and 300 K. Combustion occurs, resulting in gradual release of heat along the furnace. The stock occupies the whole of the floor of the furnace. The specific heat capacity of the gases is 1250 J/kg K. By sub-dividing the furnace into five 1 m × 1 m × 1 m zones (see Figure 6.2), determine the temperature distribution of the gases within the furnace, given the following data:

Zone	1	2	3	4	5
Heat Release (kW)	2400	1000	100	0	0
Total Absorption Coefficient (kp)	2.0	1.5	1.0	0.6	0.4

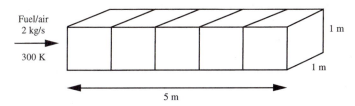

Figure 6.2 Schematic of long furnace for Example 6.1, showing five zones.

We shall assume that the gases at the inlet and outlet to the furnace are cold and black (i.e. there is no emission from the gases in the plane of the entrance and exit), and the walls and roof of the furnace are refractory and somewhat colder than the gases. Additionally, in order to simplify the calculations in this example, we will assume no heat transfer across the zone boundaries, i.e. the zone boundaries are also essentially refractory.

Solution.

The emissivity of the gases in each zone can be calculated from the mean beam length and Equation 5.11. In each zone, the gases are radiating to the six faces of the cube. So, using Table 5.1, we have a mean beam length (L_m) = $0.6 \times 1 = 0.6$ m. Equation 5.11 then gives:

Zone	1	2	3	4	5
$\varepsilon_g = 1 - e^{-kpL_m}$	0.699	0.593	0.451	0.302	0.213

An energy balance can be applied to each zone. Heat is released by combustion which is then transferred to the furnace walls and the stock by radiation and convection. (The latter will be neglected in this example.) Radiation will also be received by the gases from the furnace walls. If the rate of heat transfer from the gases is higher or lower than the rate of heat release, then any surplus/deficit results in a rise/fall in the temperature of the gases as they pass through the zone. The heat balance is therefore:

$$\begin{matrix} \text{Heat Release} \\ \text{by Combustion} \end{matrix} = \begin{matrix} \text{Radiation} \\ \text{(gas} \to \text{walls)} \end{matrix} - \begin{matrix} \text{Radiation} \\ \text{(walls} \to \text{gas)} \end{matrix} + \begin{matrix} \text{Convection} \\ \text{(gas} \to \text{walls)} \end{matrix}$$
$$+ \begin{matrix} \text{Enthalpy rise of} \\ \text{the gases} \end{matrix} \qquad (6.4)$$

In any zone, we will assume that the walls are cold compared with the gases, so the radiation received by the gases (the second term on the right-hand side) will be negligible. We shall also assume that convection is negligible, so the balance becomes, algebraically:

$$Q = A_w \varepsilon_g \sigma T_g^4 + mc_p(T_o - T_i)$$

where A_w is the surface area of wall removing heat from gases, $\dot{m}$ is the mass flow rate of gases, T_o is the outlet temperature of the gases from the zone, and T_i is the inlet temperature of the gases to the zone. We shall take T_g^4, the gas temperature in a zone, to be the arithmetic mean of the fourth powers of the inlet and outlet temperatures. The balance then becomes:

$$Q = \tfrac{1}{2}A_w\varepsilon_g\sigma\left(T_o^4 + T_i^4\right) + \dot{m}c_p(T_o - T_i) \tag{6.5}$$

From Equation 6.5, we can see in our example that if we know the inlet temperature for a given zone, then the outlet temperature is the only unknown. Thus, starting from zone 1, we can solve for T_o, albeit from a non-linear equation, and use that value as the inlet temperature for the next zone. Thus we can build up a temperature profile for the gases along the axis of the furnace.

Let us consider zone 1. Here heat is lost by radiation to the stock and to the cold front end of the furnace, with other walls and the boundary with zone 2 being refractory. Thus we have:

$$2.4 \times 10^6 = \left(\tfrac{1}{2} \times 2 \times 0.699 \times 5.67 \times 10^{-8} \times (300^4 + T_o^4)\right) \\ + (2 \times 1250 \times (T_o - 300))$$

$$T_o^4 + 6.310 \times 10^{10}\, T_o = 7.949 \times 10^{13}$$

Numerical solution gives $T_o = 1224$ K for zone 1, which can be used as T_i for zone 2. Noting that only the stock removes heat in zone 2, the heat balance for zone 2 is:

$$1.0 \times 10^6 = \left(\tfrac{1}{2} \times 1 \times 0.593 \times 5.67 \times 10^{-8} \times (1224^4 + T_o^4)\right) \\ + (2 \times 1250 \times (T_o - 1224))$$

$$T_o^4 + 1.486 \times 10^{11}\, T_o = 2.391 \times 10^{14}$$

giving $T_o = 1568$ K for zone 2. Progressing one zone at a time along the furnace gives us the temperature profile:

Zone	1	2	3	4	5
T_i	300	1224	1568	1548	1510
T_o	1224	1568	1548	1510	1463

This calculation shows that the approximation that zones are isothermal is probably fair in zones 3 to 5, but is not so good in the main combustion zone. If greater accuracy is required, then we might considered splitting the first two zones into smaller units, and repeating the calculations.

This rough calculation gives us a good idea of what is happening to the gases in the furnace. The assumptions which have been made do not allow us to determine the wall temperature. This would require us to include the second or third term in Equation 6.4, so that the wall temperature becomes an unknown in each zone. Also, if we needed to allow for heat transfer across zone boundaries, then the first term on the right-hand side would need to include radiation from the gases of the zone in question to all surfaces in the furnace, and the second term would include radiation from walls and gases of other zones. Such extra complexities are best suited to computer simulation.

7 Combined modes of heat transfer

7.1 Introduction

We have so far considered only radiative heat transfer and have neglected any effects of conduction and convection. In all practical systems there will nearly always be contributions from all three heat transfer modes. For example, in a furnace the stock will be heated by a combination of radiation and convection, and conduction is important in transferring heat from the surface to the interior of the stock. Conduction is also responsible for any heat losses through the walls of the furnace. We must therefore consider the analysis of systems where two or more mechanisms are occurring simultaneously. Computation is inevitable made more difficult when more than one mechanism operates, with a full analytical solution often leading to non-linear differential or integro-differential equations.

If only rough hand-calculation is required, such complexities are inconvenient and considerable time and effort can be saved, if, from the outset, it is recognised that one or more heat transfer mechanisms can be neglected. Thus the engineer needs some idea of when any particular mode is likely to dominate or, conversely, may be safely ignored. A few rules-of-thumb can be summarised as follows:

(1) Conduction is generally important only for heat transfer through solids. For opaque solids, conduction can be assumed to be the only mechanism, but radiation is also important in transparent or porous solids, including packed beds and fluidised beds.

(2) Radiation is the only possible mechanism through a vacuum. It also is dominant in many low pressure (sub-atmospheric) systems.

(3) Radiation dominates over convection in high temperature furnaces (> 1000 °C), unless there is strong forced convection with high gas velocities, as is found in gas-fired rapid heaters.

(4) Convection normally outweighs radiation in all low temperature systems (<~ 400 °C) in which there is forced convection.

(5) Radiation and convection are comparable from external surfaces in stagnant surroundings such as central heating radiators, heat losses from pipes.

(6) Radiation dominates over convection with very large objects, e.g. the earth, where the surface temperature on a still, cloudless night can drop below the air temperature because of radiation to the much colder outer atmosphere.

Example 6.1 introduced the idea of incorporating convective heat transfer into a radiative problem, by including a convection term in the heat balance (Equation 6.4). Convection between the gases and the walls could have been quantified in the form $h(T_g-T_w)$, where T_w is the temperature of the zone walls (assumed constant in any zone). This chapter considers two further approaches to dealing with combined modes of heat transfer. First we look at the use of heat transfer coefficients in the context of radiative heat transfer, and then we look at analytical methods for combining heat transfer modes.

7.2 The radiative heat transfer coefficient

Convective heat transfer between a surface and a fluid is usually described in terms of a convective heat transfer coefficient, h_{conv}, defined as the ratio of the convective heat flux (Q_{conv}/A) to the temperature difference between the surface and the surrounding fluid:

$$h_{conv} = \frac{Q_{conv}}{A(T_{surface} - T_{surr})} \tag{7.1}$$

In principle we can define a radiative heat transfer coefficient in exactly the same way, i.e. the ratio of the net radiative flux (Q_{rad}/A) to the difference in temperature between a hot body (the source) and a cold body (the receiver):

$$h_{rad} = \frac{Q_{rad}}{A(T_{source} - T_{receiver})} \tag{7.2}$$

In order to make use of these expressions to calculate the convective or radiative heat flux, we need to determine these heat transfer coefficients. The convective coefficient is generally found using experimental correlations involving such dimensionless groups as the Reynolds, Prandtl, Nusselt and Grashof Numbers, which are primarily functions of the fluid properties. Correlations are given in most heat transfer texts. An expression for the radiative coefficient is obtained by substituting Equation 4.11 for Q_{rad}:

$$h_{rad} = \frac{Q_{rad}}{T_{source} - T_{receiver}} = \frac{\mathcal{F}\left(\sigma T_{source}^4 - \sigma T_{receiver}^4\right)}{T_{source} - T_{receiver}}$$

$$h_{rad} = \mathcal{F}\sigma\left(T_{source}^2 + T_{receiver}^2\right)(T_{source} + T_{receiver}) \tag{7.3}$$

The radiation factor $\mathcal{F}$ can be calculated using the methods described in earlier chapters, and will depend on the geometry and the emissivities of the materials.

Use of h_{rad} is often convenient in combined mode heat transfer problems. Suppose that we need to determine the heat loss from a pipe through which a hot fluid is flowing. We can consider four heat transfer mechanisms which result in the transfer of heat from the hot fluid to the surroundings:

(i)   forced convection from the fluid to the inner pipe wall;
(ii) conduction through the pipe wall;
(iii) natural convection from the outer pipe wall to the surroundings;
(iv) radiation from the outer pipe wall to the surroundings.

Radiative effects may also be present inside the pipe. These will be neglected here, but are discussed in more detail in Section 7.3.3. Note that mechanisms (iii) and (iv) act in parallel. Both are responsible for transfer of heat from the outer wall to the surroundings; the net transfer from the outer wall will be the sum of the two contributions.

From basic principles of heat transfer, the net rate of heat loss from the hot fluid to the surroundings can be written:

$$\text{Rate of Heat Loss} = UA_2 \,(\text{LMTD}) \tag{7.4}$$

where U is the overall heat transfer coefficient from the hot fluid to the surroundings based on outer area A_2, and LMTD is the log mean temperature difference between the hot fluid in the pipe and the bulk of the fluid surrounding the pipe. For the concentric cylindrical geometry under discussion here, U can be expressed in terms of the individual heat transfer coefficients and the conductivity of the wall of the pipe:

$$\frac{1}{Ur_2} = \frac{1}{h_1 r_1} + \frac{\ln\left(\dfrac{r_2}{r_1}\right)}{k_\text{w}} + \frac{1}{r_2(h_2 + h_\text{rad})} \tag{7.5}$$

where r_1 and r_2 are the inner and outer radius of the pipe, h_1 and h_2 are the convective heat transfer coefficients at the inner and outer walls, and k_w is the thermal conductivity of the pipe wall. For thin-walled metal pipes, the conductivity term can usually be neglected. If the outer wall of the pipe is grey, then $\mathcal{F} \rightarrow \varepsilon_2$, the emissivity of the outer wall (Section 4.3). If the temperatures of the fluid and the surroundings are known, then we can calculate first the convective and radiative heat transfer coefficients, using appropriate correlations and Equation 7.3. (Note that these equations assume a constant wall temperature for the pipe; if this is not a fair assumption, then the wall will need to be divided into a number of zones, in each of which the wall can be assumed to be isothermal.) Equation 7.5 gives U and finally the heat loss is found using Equation 7.4.

7.3   Analytical methods

The requirement for constant wall temperature in the above analysis, and the possibility that surfaces may need to be sub-divided leads naturally to analytical methods, where we consider heat transfer to and from differential elements in order to build up sets of differential equations, which, hopefully, can be solved to give the required solution to the problem. This section considers a number of examples of radiation combined with conduction, with convection, and with both conduction and convection. It shows the principles of how to set up the governing equations, but does not offer methods of solution. Most of the equations require numerical solution, and, if unfamiliar with the necessary techniques, the reader should refer to appropriate mathematical texts.

7.3.1 Radiation with conduction

Radiation combined with conduction is a system which is generally confined to heat transfer through a vacuum, since the presence of any fluid at all is likely to produce convection effects in some form. It is of principal importance in space technology, where radiation is the only means of heat transfer to or from a spacecraft. We shall consider as an example heat losses by radiation from a heat exchanger fin.

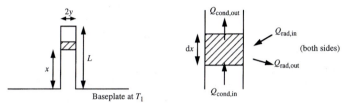

Figure 7.1 Schematic of radiating fin, showing (right) heat balance on a fin element.

Consider a single fin of length L and thickness $2y$ attached to a surface of uniform temperature T_1, which is higher than T_0 the temperature of the surroundings (see Figure 7.1). Heat is lost from the fin by conduction along the fin, followed by radiation from the fin surface. In analysing the heat losses we shall assume the following:

(1) The fin is very thin compared with its length, such that there is a temperature gradient only along the length of the fin, and not across it.
(2) If the fin is very thin, we can also assume that radiation losses from the tip of the fin (in the x-direction) are negligible compared with radiation from the sides. Algebraically this can be expressed $(\mathrm{d}T/\mathrm{d}x)_{x=L} \approx 0$.
(3) All surfaces are opaque with constant emissivity ε and absorptivity α.
(4) The fin material has constant thermal conductivity k.

Consider a small element of fin thickness $\mathrm{d}x$ (Figure 7.1). A heat balance of conduction and radiation effects gives:

  Conduction into element + Radiation received =

                 Conduction out + Leaving Radiation

Algebraically:

$$-k(2y)\frac{\mathrm{d}T}{\mathrm{d}x} + G(2\,\mathrm{d}x) = -k(2y)\frac{\mathrm{d}}{\mathrm{d}x}\left(T + \frac{\mathrm{d}T}{\mathrm{d}x}\mathrm{d}x\right) + J(2\,\mathrm{d}x) \qquad (7.6)$$

Note the allowance for radiation emitted/received from both sides of the element. The incident flux will consist of radiation received from the surroundings and the baseplate of the fin. We can introduce the definition of radiosity and then simplify Equation 7.6 to give:

$$-ky\frac{\mathrm{d}^2 T}{\mathrm{d}x^2} = \varepsilon\sigma T^4 - \alpha G \qquad (7.7)$$

to which we can apply two boundary conditions: $T = T_1$ at $x = 0$, and $(\mathrm{d}T/\mathrm{d}x) = 0$ at $x = L$.

Clearly we have a non-linear differential equation, which will in general require a numerical solution to give the temperature distribution as a function of x. Having obtained the temperature distribution, the rate of heat loss from the fin is then given by a heat balance over the whole fin:

$$Q = -k(2y)\frac{dT}{dx}\bigg|_{x=0} = 2\varepsilon\sigma\int_0^L T^4\,dx - 2\alpha\int_0^L G\,dx \qquad (7.8)$$

The most difficult term to deal with in Equations 7.7 and 7.8 is the final one, since the incident flux, G, depends on the extent to which the element can see the surroundings and the baseplate. This will vary with x, because elements closer to the baseplate will receive proportionally more radiation from that surface. We will consider three cases of increasing complexity:

(1) *Incident radiation is negligible.* We will first assume that $G \approx 0$, in which case Equation 7.7 becomes:

$$-ky\frac{d^2T}{dx^2} = \varepsilon\sigma T^4 \qquad (7.9)$$

It is helpful to introduce a dimensionless distance $X\ (= x/L)$ and a dimensionless temperature $\theta\ (= T/T_1)$, which lead to:

$$\frac{dT}{dx} = \frac{T_1}{L}\cdot\frac{d\theta}{dX} \quad \text{and} \quad \frac{d^2T}{dx^2} = \frac{T_1}{L^2}\cdot\frac{d^2\theta}{dX^2}$$

Subsitution into Equation 7.9 gives:

$$\frac{d^2\theta}{dX^2} - N\theta^4 = 0 \qquad (7.10)$$

where $N = \dfrac{\varepsilon\sigma T_1^3 L^2}{ky}$ and the boundary conditions become $\theta = 1$ at $X = 0$, and $(d\theta/dX) = 0$ at $X = L$. N is a measure of the relative contributions of radiation and conduction, and is therefore a form of Biot Number.

(2) *Incident radiation comes from the surroundings only.* If incident radiation from the surroundings cannot be neglected, then we can assume that $G = \sigma T_o^4$, noting that the surroundings always appear black to a sufficiently small object (Section 4.3). Equation 7.7 therefore becomes:

$$-ky\frac{d^2T}{dx^2} = \varepsilon\sigma T^4 - \alpha\sigma T_o^4$$

Using the same dimensionless groups as before, together with $\theta_o = T_o/T_1$, this equation becomes

$$\frac{d^2\theta}{dX^2} - N\left(\theta^4 - \frac{\alpha\theta_o^4}{\varepsilon}\right) = 0 \qquad (7.11)$$

which must be solved numerically. Note that this equation applies for any non-grey surface; assuming a grey fin gives very little further simplification.

(3) *Incident radiation comes from surfaces and surroundings.* The fin shown in Figure 7.1 will receive radiation not only from the surroundings, but also from the surface supporting the fin (see Figure 7.2). The rate at which radiation from the surface impinges on the fin element is given by:

$$(\text{Transfer Rate})_{1 \to dx} = J_1 A_1 F_{1 \to dx} = J_1 F_{dx \to 1} dx$$

where $F_{dx \to 1}$ is the view factor from the fin element to the baseplate, and J_1 is the radiosity of the baseplate. Similarly the rate at which radiation from surroundings impinges on the fin element is given by:

$$(\text{Transfer Rate})_{surrs \to dx} = \sigma T_0^4 F_{dx \to surrs} dx = \sigma T_0^4 (1 - F_{dx \to 1}) dx$$

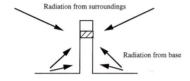

Radiation from surroundings

Radiation from base

Figure 7.2 Schematic showing radiation impinging on fin.

Equation 7.7 then becomes:

$$-ky\frac{d^2 T}{dx^2} = \varepsilon \sigma T^4 - \alpha \sigma T_0^4 (1 - F_{dx \to 1}) - \alpha J_1 F_{dx \to 1}$$

Note that $F_{dx \to 1}$ is a function of x, since the view factor from the fin element to the baseplate depends on the distance of the element from the base. An expression for $F_{dx \to 1}$ as a function of x is therefore required before a solution can be obtained to the whole equation.

All of the above analysis has assumed a single fin. Most heat exchange systems will consist of a series of fins, in which case a full analysis must also include heat transfer between adjacent fins. Further details are given in more comprehensive tests, such as Siegel and Howell.

One final aspect of fins is the concept of fin effectiveness. If we recall that the purpose of a fin is to dissipate heat, then we can, in principle, compare the radiative loss from our fin with the maximum possible. The maximum would be the radiative loss if the whole fin were at the same temperature as the baseplate. Thus the fin effectiveness (η) is defined by:

$$\eta = \frac{\text{Actual Heat Transfer from fin}}{\text{Heat Transfer if whole fin is at } T_1}$$

$$= \frac{\int_0^L \varepsilon \sigma T^4 (2dx)}{\varepsilon \sigma T_1^4 (2L)} = \int_0^1 \theta^4 \, dX \qquad (\text{if } \varepsilon \neq f(T))$$

So, if θ can be found from solving the appropriate differential equation, then a further integration will give the fin effectiveness. Clearly from the definition $0 < \theta < 1$. A high fin effectiveness is obtained for low values of N, i.e. when conduction dominates such that there is very little temperature gradient and the fin is essentially isothermal at T_1.

7.3.2   Radiation with conduction and convection

The previous section can be extended to give an example of all three modes of heat transfer, if we allow heat to be lost by the fin by a combination of radiation and convection. If the fin is placed in a stagnant fluid, then convection effects can usually be neglected, but if the fluid is moving, then forced convection effects are likely to be significant. We will consider the same geometry as before, but assume that the surrounding fluid to be at temperature T_f. The problem, as before, is to find the temperature distribution along the fin, and hence the rate of heat loss. This is achieved by a simple modification to the energy balance, incorporating an additional heat transfer term to account for convection between the fin and the surrounding fluid. Thus Equation 7.7 becomes:

$$-ky\frac{\mathrm{d}^2T}{\mathrm{d}x^2} = \varepsilon\sigma T^4 + h(T - T_f) - \alpha G$$

where h is the convective heat transfer coefficient. Use of an appropriate expression for the incident flux will lead to a non-linear equation which may be solved numerically, as before.

7.3.3   Radiation with convection

As an example of radiation and convection, we will extend the pipe flow example in Section 7.2 to include radiation effects inside the pipe. Consider flow of a hot fluid within a colder pipe. The fluid will cool down as it passes along the pipe. In Section 7.2 we assumed that the wall temperature remained constant; in reality the wall temperature will also decrease in the direction of flow. Suppose that we wish to determine the temperature profile of both the fluid and the wall as a function of distance along the pipe, taking into account convective heat transfer between the fluid and the pipe wall, and radiation effects within the pipe. The latter may be significant if one end of the pipe is significantly hotter than the other.

Suppose that we have a thin-walled pipe of length L and diameter D, through which a fluid flows at mass flowrate $\dot{m}$, with inlet temperature T_{g1} and exit temperature T_{g2} (Figure 7.3). Consider a volume element $\mathrm{d}V$ of length $\mathrm{d}x$,

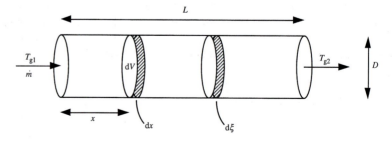

Figure 7.3   Schematic for heat transfer for gas flow in a pipe.

where the gas temperature is T_g, the wall temperature is T_w, and these temperatures change by dT_g and dT_w respectively. We shall assume here only the case of a transparent fluid, and a pipe wall which is grey and of uniform emissivity ε.

Since the fluid is non-participating, it loses heat only by convective heat transfer to the wall. An energy balance on the gas in volume dV therefore gives:

$$\text{Rate of heat transfer} = \dot{m}c_p dT_g = h_1(T_w - T_g)\pi D\, dx$$

noting that dT_g is positive when T_w is greater than T_g. This leads to the differential equation

$$\frac{dT_g}{dx} = \frac{h_1 \pi D}{\dot{m}c_p}(T_w - T_g) \tag{7.12}$$

with the boundary condition $T_g = T_{g1}$ at $x = 0$, and $T_g = T_{g2}$ at $x = L$. An energy balance can also be carried out on the surface of the pipe in volume dV, in terms of convective and radiative fluxes:

$$\text{Net Flux to wall} = \text{Convection (gas} \rightarrow \text{wall)} + \text{Incident Radiation}$$

$$- \text{Leaving Radiation}$$

$$q = h_1(T_g - T_w) + G - J$$

$$= h_1(T_g - T_w) + \alpha G - \varepsilon\sigma T_w^4 \tag{7.13}$$

The incident flux has three components: radiation received from the inlet end, radiation received from the outlet end, and radiation received from the pipe walls. (If the fluid were participating, then we would also include radiation from the fluid.) G is therefore a function of x, and we must derive an appropriate expression in terms of temperatures. Although the fluid is non-participating, we shall assume that radiation is received from the pipe ends at their respective temperatures, and that the pipe ends are black. The three components of the incident radiation can be expressed algebraically as:

$$G = \sigma T_1^4 F_{dx \rightarrow 1} + \sigma T_2^4 F_{dx \rightarrow 2} + \int_0^L J F_{dx \rightarrow d\xi} d\xi$$

where J, the radiosity of the wall is a function of x. The last term, the radiation received from the pipe walls, is determined by considering heat transfer between the wall element dx and another wall element $d\xi$, and then integrating over all the $d\xi$ (see Figure 7.3). Analytical expressions for the three view factors are available in tables. Substitution into Equation 7.13 gives:

$$q = h_1(T_g - T_w) + \varepsilon\left(\sigma T_1^4 F_{dx \rightarrow 1} + \sigma T_2^4 F_{dx \rightarrow 2} + \int_0^L J F_{dx \rightarrow d\xi} d\xi - \sigma T_w^4\right)$$

$$\tag{7.14}$$

Continuity of heat flow requires the rate of transfer of heat at the outer wall is same as that at the inner wall. Therefore we can also write:

$$q = h_2(T_w - T_o) + \varepsilon(\sigma T_w^4 - \sigma T_o^4) \qquad (7.15)$$

which can be combined with Equation 7.14 to give a second equation in terms of T_g and T_w. This together with Equation 7.12 can be solved numerically to give the two required temperature profiles.

8 Measurement of temperature

8.1 Introduction

Techniques for the measurement of temperature may be divided into two general types. Intrusive methods, such as thermometers and thermocouples, involve the introduction into the system of a probe or detector, whose output signal gives an indication of the temperature. Very often, this temperature, which is necessarily that of the probe, is different from the temperature of the material being measured. Only a full heat transfer analysis will lead us to the required temperature. Non-intrusive methods, such as pyrometers, radiometers, and modern laser-based systems, are normally used where the temperatures are too high or the environment too hostile for a probe. The detector is placed remote from the system, but in a position where emitted radiation can be "seen". The intensity of the radiation is dependent on the temperature, so data analysis can convert the radiation signal into a temperature.

In this chapter we shall consider methods of temperature measurement which are based on the emission or absorption of thermal radiation. Thus spectroscopic methods based on the detection of radiation arising from molecular electronic or vibration transitions, or by simple scattering are not included here. All the methods discussed require some form of heat transfer analysis in order to extract correct temperature information. We shall see that the analyses normally involve a combination of all three heat transfer mechanisms, so it is appropriate to consider temperature measurement at this point.

8.2 Non-intrusive techniques

Radiometry and pyrometry essentially involve an instrument which converts the incident radiant energy into an electrical output signal. There are three main parts to any system:

(1) A detecting element, which is sensitive to a particular wavelength or range of wavelengths, and converts the radiant energy to an electrical output.
(2) An optical system to focus the incident radiation onto the detector; this may include a filter which transmits only the wavelength(s) of interest.
(3) An electronics system which displays the output from the detector; this may be a straightforward voltage output, or may include processing circuitry which enables temperature to be displayed.

A complete review of detection systems is outside the scope of this book, but we will concentrate on some basic methods of temperature measurement.

8.2.1 Radiometer

Arguably the most basic non-intrusive technique is the radiometer, which is best suited for the measurement of surface temperature. It consists of a single thermocouple or a thermopile (i.e. a number of thermocouples connected in series), which is heated by incident radiation falling upon it. Radiometers are broadband detectors, in that no filters are used, and radiation of all wavelengths lands on the sensor. They are sometimes termed thermal detectors, as they rely on the heating effect of the radiation incident upon them. The thermocouple junctions are normally coated with platinum black or carbon black in order to raise their absorptivity to very close to 1. The incident radiant flux determines the temperature of the thermocouple, and hence the electrical output from the radiometer. The incident flux will depend not only on the temperature of the radiation source, but also on how close the radiometer is to the source. A radiation analysis is therefore required to relate the radiometer output to the temperature of the source.

Figure 8.1 shows schematically the arrangement for a radiometer (surface 2) measuring the surface temperature of a hot object (surface 1).

If the medium between the surface and the radiometer is non-participating, then the rate of radiative heat transfer, Q_{12} from surface 1 to the radiometer is:

$$Q_{12} = \varepsilon_1 A_1 F_{12}\left(\sigma T_1^4 - \sigma T_2^4\right) = \varepsilon_1 A_2 F_{21}(\sigma T_1^4 - \sigma T_2^4)$$

In normal operation of a radiometer, $T_2 << T_1$, in which case this equation can be simplified to give the incident flux $G_2(= Q_{12}/A_2)$:

$$G_2 = \varepsilon_1 F_{21} \sigma T_1^4 \tag{8.1}$$

Radiometers are generally installed to give a simple voltage output on a voltmeter. A calibration factor (K) in Watts per unit radiometer surface area per Volt output is supplied by manufacturers, which gives a direct conversion of the voltage output (V) into the incident flux. Thus we can also write:

$$G_2 = KV \tag{8.2}$$

Combining Equations 8.1 and 8.2 gives an expression for the temperature of surface 1:

$$T_1 = \left(\frac{KV}{\sigma \varepsilon_1 F_{21}}\right)^{0.25} \tag{8.3}$$

So, in order to convert the voltage output to a surface temperature, we need to know the emissivity of surface 1, and must be able to determine the view

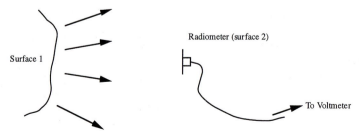

Figure 8.1 Schematic arrangement of a radiometer measuring surface temperature.

factor from the detector to surface 1. The latter requires some knowledge of the geometry of the detector; manufacturers will supply the surface area of the detector, and the cone angle of view of the sensing element. Because there is no filter arrangement on the detector, it is appropriate to use the total emissivity of the hot surface.

If the medium between the surface and the detector is participating, then some allowance needs to made in the radiation analysis, using the methods outlined in earlier chapters. Clearly the absorptivity and emissivity of the medium must be known (or calculable) in order to determine the surface temperature.

Although straightforward both in principle and in use, the radiometer suffers from the major disadvantage that it receives radiation from all surfaces within its field of view, and not just the surface of interest. The detector must therefore be positioned carefully so that it "sees" only the surface of interest. Another problem is that it will give only an average temperature of the surface, so should not be used where any spatial resolution of surface temperature is required.

8.2.2 Disappearing filament pyrometer

The basis of the disappearing filament pyrometer is the matching of radiation from a variable, calibrated source with that from the surface of interest. Figure 8.2 shows a typical arrangement.

The instrument consists of a series of lenses, a filament, and a filter, which in most instruments allows transmission of radiation of wavelength 650 nm in the red part of the visible spectrum. The surface is viewed through the eye-piece and focussed onto the plane of the filament. The current through the filament is then adjusted, until the brightness of filament matches that of the surface. Under these circumstances, the filament merges (or disappears) into the image of the surface and the flux from the surface and from the filament at 650 nm can be assumed to be equal. Note that this does not mean that the temperatures of the surface and the filament are equal, unless their emissivities are also equal.

In commercial instruments the filament is normally calibrated against the temperature of a black body source. Thus the temperature indicator on the ins-trument displays the black body temperature, not the real temperature of the surface. We must therefore calculate the real temperature of the surface from the measured black body temperature. Because of the narrow bandwidth filter, we use Planck's Law for the monochromatic emissive power, rather than

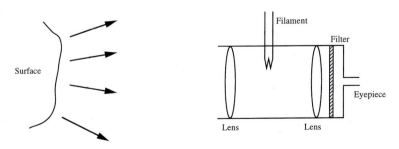

Figure 8.2 Schematic of a disappearing filament pyrometer.

the Stefan–Boltzmann Law. Further, because we are dealing with visible radiation at modest temperatures, it valid to use Wien's Approximation to Planck's Law (Equation 1.14). The emissive flux at wavelength λ from the surface at temperature T is given by

$$W_\lambda = \varepsilon_\lambda \frac{2\pi hc^2}{\lambda^5} \exp\left(-\frac{hc}{\lambda kT}\right) \tag{8.4}$$

where ε_λ is the monochromatic emissivity of the surface at wavelength λ. When the filament disappears, this flux matches that from a black body at the temperature T_b displayed on the instrument. So, we also have

$$W_{b\lambda} = \frac{2\pi hc^2}{\lambda^5} \exp\left(-\frac{hc}{\lambda kT_b}\right) \tag{8.5}$$

Equating the fluxes in Equations 8.4 and 8.5 and rearranging gives

$$\frac{1}{T} = \frac{\lambda k}{hc} \ln \varepsilon_\lambda + \frac{1}{T_b} \tag{8.6}$$

Thus the true surface temperature can be calculated if the monochromatic emissivity of the surface is known. Equation 8.6 shows that the displayed temperature must always be an underestimate of the real temperature. For high emissivity surfaces (> 0.8), the difference between the indicated and real temperature is quite small (under 1%) and to a first approximation the indicated temperature can be taken to be the true temperature; for low emissivity surfaces differences can be quite large (up to 10%), so large errors can result if the correction is not made. (N.B. Modern instruments can incorporate into their electronics the analysis which converts black body temperature into true temperature, if the user enters a value for the emissivity of the surface.)

If the emissivity of the surface is not known, then a pyrometer with filters of two different wavelengths λ_1 and λ_2 can be used. The surface is viewed with each filter separately, and the pyrometer displays different black body temperatures, T_{b1} and T_{b2} for each filter. We can then write

$$\frac{1}{T} = \frac{\lambda_1 k}{hc} \ln \varepsilon_{\lambda 1} + \frac{1}{T_{b1}}$$

$$\frac{1}{T} = \frac{\lambda_2 k}{hc} \ln \varepsilon_{\lambda 2} + \frac{1}{T_{b2}}$$

where $\varepsilon_{\lambda 1}$ and $\varepsilon_{\lambda 2}$ are the monochromatic emissivities at wavelengths λ_1 and λ_2, respectively. If we can assume that the transmission wavelengths of the two filters are sufficiently close that $\varepsilon_{\lambda 1} = \varepsilon_{\lambda 2}$, the equations can be rearranged to give expressions for the true temperature and, if required, the emissivity:

$$T = \frac{\lambda_1 - \lambda_2}{\left(\dfrac{\lambda_1}{T_{b2}} - \dfrac{\lambda_2}{T_{b1}}\right)}$$

$$\ln \varepsilon_\lambda = \frac{hc}{k(\lambda_2 - \lambda_1)} \left(\frac{1}{T_{b1}} - \frac{1}{T_{b2}}\right)$$

8.3 Probe techniques

Probe techniques are generally based on instruments such as thermometers and thermocouples. The thermometer is of little engineering use, other than for measurements around ambient temperature. The thermocouple is perhaps the most common industrial technique for measurement of above-ambient temperatures, and is available in a variety of metal combinations for different applications. The probe must be in intimate contact with the material whose temperature is being measured. Thus the probe must be fixed to the particular surface or placed in the fluid stream. After coming into contact with the material the probe will come to some equilibrium temperature, which need not be the same temperature as the material we are trying to measure. This leads to measurement errors, which are an inherent feature of the technique, and not errors in calibration. It is important that suitable correction procedures be applied in order to determine the true temperature, or that steps be taken to minimise that initial error.

8.3.1 Thermocouples and thermometers

Consider a thermocouple or thermometer placed in a hot gas stream which is flowing along a pipe with relatively cool walls. The sensor (i.e. the thermocouple bead or thermometer bulb) will be subject to a number of heat transfer mechanisms which will determine its final equilibrium temperature:

(1) Convection heat transfer between the gas stream and the surface of the sensor.
(2) Radiation exchange between the sensor and the surrounding surfaces (there may also be radiation effects between the sensor and the gas if the gas is an absorbing/emitting medium).
(3) Conduction of heat along the thermocouple wires or thermometer body from the sensor to the cooler surroundings.

The conduction and radiation losses combine to ensure that the equilibrium temperature of the sensor is always less the true gas temperature. (If we have a cold fluid flow with relatively warmer surroundings, then the sensor will attain a temperature higher than the true temperature.) Conduction effects can be minimised by using thermocouple wires of very small diameter, or a thermometer body made from thermally insulating material. Radiation effects cannot usually be ignored, and must be corrected for by considering an energy balance on the surface of the sensor.

Consider a fluid of temperature T_f in which is immersed a temperature sensor of emissivity ε_c, which at equilibrium attains a temperature T_c (see Figure 8.3). Suppose that the fluid flows along a pipe whose walls are at temperature T_w (lower than T_f). If conduction effects are negligible, then an energy balance on the sensor gives:

Convective Transfer (Fluid $\rightarrow$ Sensor) = Radiative Transfer (Sensor $\rightarrow$ Walls)

If we assume that all surfaces are grey, and that the fluid is non-participating, the balance may be written:

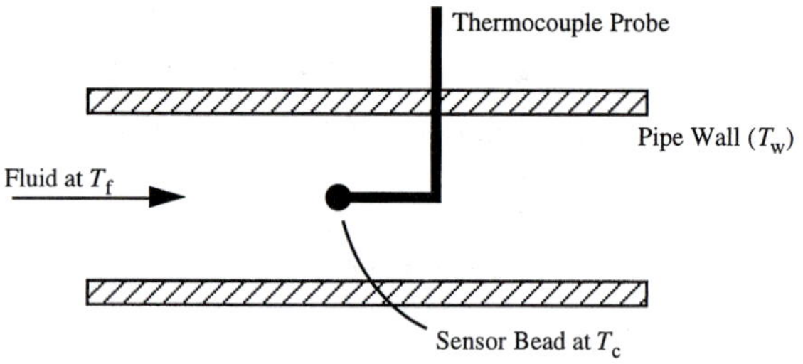

Figure 8.3   Thermocouple probe for measurement of fluid temperature.

$$h_{fc}(T_f - T_c) = \mathcal{F}_{cw}(\sigma T_c^4 - \sigma T_w^4)$$

where h_{fc} is the convective heat transfer coefficient between the fluid and   the sensor, and $\mathcal{F}_{cw}$ is the grey radiation factor for radiative transfer between the sensor and the walls. Since the sensor is usually very small compared with the enclosure in which it is sited, $\mathcal{F}_{cw} \approx \varepsilon_c$ and the equation becomes, after rearrangement:

$$T_f = T_c + \frac{\varepsilon_c \sigma}{h_{fc}}(T_c^4 - T_w^4) \qquad (8.7)$$

Thus, in order to determine the true temperature, we must know the temperature of the surrounding surfaces and the emissivity of the sensor surface. We must also be able to calculate (or estimate) a value for the convective heat transfer coefficient.

Ideally we would like our sensor to register the true gas temperature, so that no correction is required. Equation 8.7 shows that the error $(T_f - T_c)$ is large either if the temperatures of the sensor and its surroundings are very different (giving large radiation losses), or if the convective heat transfer coefficient is very small (e.g. where fluid velocities are low). (The latter explains why garden thermometers can give erroneously low readings on still, cold nights, and why thermometers whose bulb is placed in direct sunlight indicate a temperature which is much higher than the air temperature.) We therefore have two methods, in principle, by which the error can be reduced: either decrease the radiative exchange or improve convective heat transfer to the sensor. Both of these methods are used in practice and form the basis of the shielded thermocouple and the suction pyrometer.

8.3.2   Shielded thermocouples

We have seen that high radiation errors with thermocouples are caused by the bead being able to "see" surroundings which are at a much lower or higher temperature than the fluid in which the couple is placed. Radiation errors can be greatly reduced if the thermocouple bead is *shielded*, i.e. surrounded by a thin shield of material placed a short distance from the bead, such that the thermocouple is still in contact with the fluid, but can see only the shield and not the surroundings. Figure 8.4 shows a typical shield arrangement for a butt-welded thermocouple in a pipe flow.

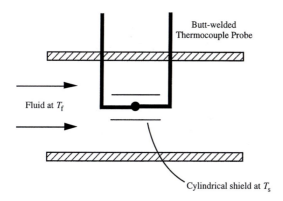

Butt-welded
Thermocouple Probe

Fluid at T_f

Cylindrical shield at T_s

Figure 8.4   Shielded thermocouple probe.

In this arrangement the thermocouple bead can see only the shield, which, being completely immersed in the fluid flow, should be much closer to the fluid temperature than the walls. Thus radiation effects should be reduced. A heat transfer analysis produces the appropriate equation to convert the bead temperature to the fluid temperature.

Radiative Transfer (Couple → Shield) $= Q_{cs} = A_c \mathcal{F}_{cs}(\sigma T_c^4 - \sigma T_s^4)$ (8.8)

Radiative Transfer (Shield → Walls) $= Q_{sw} = A_s \mathcal{F}_{sw}(\sigma T_s^4 - \sigma T_w^4)$ (8.9)

Convective Transfer (Fluid → Couple) $= Q_{fc} = h_{fc} A_c (T_f - T_c)$ (8.10)

Convective Transfer (Fluid → Shield) $= Q_{fs} = 2 h_{fs} A_s (T_f - T_s)$ (8.11)

Where h_{fc} and h_{fs} are convective heat transfer coefficients at the surface of the couple and the shield, respectively. (Note the factor of 2 in Equation 8.11, showing that convective heat transfer between the fluid and the shield occurs at both the inner and outer surfaces of the shield.) There is no radiation term for transfer between the couple and the walls, as we are assuming that the shield is sufficiently long that the couple bead cannot see the walls.

When the system has reached steady state, then, assuming no conductive losses:

Convective Transfer (Fluid → Couple) = Radiative Transfer
                                              (Couple → Shield)
Convective Transfer (Fluid → Shield) = Radiative Transfer
                                              (Shield → Couple and Walls)

or, algebraically:

$$Q_{fc} = Q_{cs} \qquad Q_{fs} = Q_{sw} - Q_{cs}$$

$$h_{fc}(T_f - T_c) = \mathcal{F}_{cs}(\sigma T_c^4 - \sigma T_s^4) \qquad (8.12)$$

$$2h_{fs}(T_f - T_s) = \mathcal{F}_{sw}(\sigma T_s^4 - \sigma T_w^4) - \frac{A_c \mathcal{F}_{cs}}{A_s}(\sigma T_c^4 - \sigma T_s^4) \qquad (8.13)$$

The grey radiation factors can normally be calculated from the system geometry and the heat transfer coefficients from the physical properties of the fluid. Equations 8.12 and 8.13 then reduce to two simultaneous equations in T_s and the required T_f.

Example 8.1.
A thermocouple of 2 mm bead diameter is placed in a pipe of 200 mm internal diameter, which is transporting a non-participating gas whose temperature is 800 K. The pipe wall temperature is 500 K, the emissivity of the couple is 0.8, and the convective heat transfer coefficient is 100 W/m²K. Calculate the expected error in the temperature measurement. How does this error change if the thermocouple if fitted with a shield of diameter 10 mm and emissivity 0.4?

Solution.
The temperature displayed by the unshielded thermocouple is governed by Equation 8.7, given that the thermocouple bead is small compared with the diameter of the pipe. From Equation 8.7:

$$\frac{\varepsilon_c \sigma}{h_{fc}} T_c^4 + T_c = \frac{\varepsilon_c \sigma}{h_{fc}} T_w^4 + T_f$$

$$4.536 \times 10^{-10}\, T_c^4 + T_c - 828.4 = 0$$

Numerical solution gives $T_c = 712$ K, which represents a measurement error of nearly 90 K.

For the shielded thermocouple we will assume that the couple-shield geometry can be approximated by concentric cylinders, in which case the grey radiation factor is given by Equation 4.13:

$$\mathcal{F}_{cs} = \frac{1}{\varepsilon_c^{-1} + \dfrac{A_c}{As}(\varepsilon_s^{-1} - 1)} = \frac{1}{\varepsilon_c^{-1} + \dfrac{d_c}{d_s}(\varepsilon_s^{-1} - 1)}$$

from which $\mathcal{F}_{cs} = 0.645$. The radiation shield is small compared with the pipe diameter, so we can take $\mathcal{F}_{cw} \approx \varepsilon_c = 0.4$. (The more accurate equation gives 0.397, so this approximation is indeed valid.) If we assume that the heat transfer coefficients at the couple and shield surfaces are the same, then equations 8.12 and 8.13 give:

$$3.657 \times 10^{-8} T_c^4 + 100\, T_c - 3.657 \times 10^{-8} T_s^4 = 80\,000$$

$$7.314 \times 10^{-9} T_c^4 - 2.999 \times 10^{-8} T_s^4 - 200\, T_s + 161418 = 0$$

Numerical solution gives a value of around $T_c = 788$ K and $T_s = 769$ K, i.e. the measurement error has decreased from nearly 90 K to just 12 K. Unless greater accuracy is required it is usually acceptable to assume that the reading from a thermocouple fitted with a radiation shield is equal to the true fluid temperature.

8.3.3 Suction pyrometer
The effect of convective heat transfer on the surface of the thermocouple is clearly shown by repeating the last example with different values of h_{fc}. For

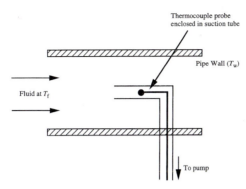

Figure 8.5 A typical layout for a suction pyrometer.

instance, increasing h_{fc} from 100 to 1000 W/m^{2}K, increases T_c for the unshielded thermocouple from 712 K to 785 K. If h_{fc} is 2000 W/m^{2}K, then T_c is 793 K, i.e. only 7 K below the true temperature. Thus, an additional method of reducing measurement error is to increase convective heat transfer from the fluid to the thermocouple bead. This forms the basis of the suction pyrometer (see Figure 8.5).

A normal thermocouple is placed inside a narrow-bore tube, which is connected to a suction pump. When the pump is turned on the fluid is sucked through the tube over the thermocouple bead with a much higher velocity than the flow of the fluid in the system. Thus the convective heat transfer coefficient at the surface of the thermocouple is greatly increased. Placing the bead inside the tube also provides some shielding of the thermocouple, which reduces radiation effects. The combination of increased convection and reduced radiation losses means that the suction pyrometer gives a temperature which, to engineering tolerances, can often be assumed to be the true temperature of the fluid. (Some correction might still be necessary, depending on the design used.) Suction pyrometers are used particularly in the analysis of combustion gases from furnaces, boilers, and other combustion systems, where the calculation of measurement errors is particularly complex because of emission/absorption by the combustion products. The suction pyrometer can therefore provide a simple, yet reliable, method for a more accurate determination of the gas temperature.

9 Radiation from flames

9.1 Introduction

The final chapter deals with radiation heat transfer in combustion system, in particular the combustion gases present in such systems. The main complication which can arise is that combustion gases are often laden with particles, which can play a significant rôle in the heat transfer processes. It is the particles present which give a flame its luminosity, and this luminosity is a major contributor to heat transfer from flame gases to the surroundings. Understanding the heat transfer processes involved is therefore of great importance in the development of the modelling of high temperature systems.

9.2 Non-luminous gases

The gaseous phase in combustion gases will contain a mixture of carbon dioxide, water vapour, unused oxygen, nitrogen, and unburnt fuel. There will also be traces of carbon monoxide and nitrogen oxides, but these small concentrations will have a negligible effect on radiation. The analysis of these combustion gases will be the same as for any absorbing/emitting gas mixture. If the gases are well mixed and isothermal we have a standard gas problem; if the gases are not isothermal we can use a zone method, with a sufficiently large number of subdivisions for each zone to be considered isothermal. Methods for solution have been given in earlier chapters.

9.3 Luminous gases

9.3.1 General

The luminosity (i.e. the yellow coloration) in combustion gases is due to glowing solid particles, which come principally from three sources:

(1) Solid fuel particles, which are usually injected into a furnace with a particle size in the range 10 to 200 μm. The particle size does not remain constant as the particle burns: there is some swelling during the devolatilisation stage, followed by a steady decrease in size until either complete burnout occurs, or a small residual ash particle is formed. Ash particles will continue to radiate after the fuel has been consumed, so the ash content of the coal has an important bearing on the radiation characteristics of the gases.

(2) Fuel oils, particularly the heavier ones containing polycyclic aromatic hydrocarbons, have a tendency to carbonise to form solid or semi-solid particles when sprayed into a hot combustion chamber. These particles

have similar characteristics to burning coal particles and their radiation properties are very similar.

(3)   Soot, which is formed in fuel-rich hydrocarbon flames with particle sizes in the range 0.005 to 0.15 μm. Soot is essentially a polymeric hydrocarbon material formed by condensation and dehydrogenation of heavy gaseous hydrocarbons and subsequent particle growth.

Note that because this radiation comes from a solid phase, it is a continuous emission over all wavelengths, and not band emission like gases. Consequently radiative transfer from solid particles is normally much greater than from gases. The latter can usually be neglected in engineering calculations.

Radiation passing through a particle-laden medium can be absorbed or scattered as a result of the interaction between radiation and particles. The theory of scattering and absorption is described fully by Maxwell's Equations. Solution of these equations is rather complex and limiting cases or simplifications are usually sought in order to make calculations more tractable for certain ranges of particle size. We find that the radiation properties of particles are crucially dependent on the refractive index of the particles and the particle size relative to the wavelength of the radiation. The latter is usually expressed as the parameter ($\pi d/\lambda$), where d is the particle diameter and λ the wavelength of the radiation. Three important regimes can be identified:

$\pi d/\lambda$	
<0.3	Rayleigh Scattering
0.3–5	Mie Scattering
>5	Reflective Scattering

The thermal radiation of engineering interest is in the visible and infra-red regions of the electromagnetic spectrum, in a range of approximately 0.5 to 10 μm. Thus fuel particles are generally of a diameter much greater than the wavelength of thermal radiation, in which case scattering is principally by reflection at the particle surface. Soot particles are generally smaller than the wavelength of thermal radiation, and fall within the regime of Rayleigh Scattering.

9.3.2 Large particles ($d > 2\lambda$)

The following analysis can be used for particles whose diameter is much larger than typical wavelengths of thermal radiation. It is therefore applicable to coal flames, oil droplets, and other solid fuels. We shall assume that the particles are black, i.e. the absorptivity and hence the emissivity are both 1. Because all scattering is reflective, this also means that there is negligible scattering of radiation from the particles in this size range. Further we shall assume that the particles are widely spaced, such that the volume fraction of particles is much less than 1.

The basis of the analysis is that a beam of radiation entering the enclosure containing the particle-laden gases either impinges on a particle and is absorbed, or passes through the gases untouched. The particles therefore simply block the radiation and prevent it from passing through. The radiation absorbed will be proportional to the projected area of particles presented to the incoming radiation. We can therefore write:

$$\frac{\text{Projected area of particles per}}{\text{unit volume of suspension}} = \frac{\text{Particle volume}}{\text{Fraction}} \times \frac{\text{Projected area of all particles}}{\text{Volume of all particles}}$$

$$K = f_v A_v \tag{9.1}$$

For a suspension of N spherical particles of diameter d,

$$A_v = \frac{N\left(\dfrac{\pi d^2}{4}\right)}{N\left(\dfrac{\pi d^3}{6}\right)} = \frac{3}{2d}$$

Substitution into Equation 9.1 gives

$$K = \frac{3f_v}{2d} \tag{9.2}$$

K represents a form of absorption coefficient which can be substituted for $k_\lambda p$ in Equations 5.2 to 5.4 to yield expressions for the absorptivity and hence emissivity of the suspension:

$$\frac{I_L}{I_0} = \tau = e^{-3f_v L/2d} \tag{9.3}$$

$$\varepsilon = \alpha = 1 - e^{-3f_v L/2d} \tag{9.4}$$

The volume fraction of particles can be estimated from the flowrates of fuel and air. If m_f kg/s of solid fuel and m_a kg/s of air are fed to a furnace, then

$$f_v = \frac{\text{Fuel Volume}}{\text{Total Volume}} \approx \frac{(m_f/\rho_f)}{((m_f + m_a)/\rho_p)} \tag{9.5}$$

where ρ_f and ρ_p are the densities of the solid fuel particles and the combustion products, respectively. Note that we approximate the total volume to the volume of all combustion products. Note also that Equation 9.4 predicts that the emissivity of coal flames increases as the combustor becomes larger, as particles become smaller, and as the pressure is increases (via its effect on gas density).

Example 9.1.
A pulverised coal of density 1300 kg/m^3 and mean particle diameter of 50 μm is burnt stoichiometrically in air at 1600 °C, 1 bar in a cylindrical combustor which has 0.5 m internal radius and is 2 m long. Calculate the emissivity of the combustion gases, assuming that the combustion gases have a mean Relative Molecular Mass of 30 kg/kmol, and that stoichiometric combustion of the coal requires 12 kg of air per kg of coal. How does the emissivity change if a chamber ten times larger ($r = 5$ m, $L = 20$ m) is used?

Solution.
We first calculate f_v using Equation 9.5 and then substitute the result into Equation 9.4 to give the emissivity. The problem does not give flowrates, but

we do know the fuel/air ratio. Thus we will take as a basis for the calculation a fuel flowrate of 1 kg/s and an air flowrate of 12 kg/s.

The mean beam length for the cylindrical chamber is given in Table 5.1:

$$L_{\mathrm{m}} = \frac{1.8rL}{L+r} = \frac{1.8 \times 0.5 \times 2}{2 + 0.5} = 0.72 \text{ m}$$

From the ideal gas law:

$$\rho_{\mathrm{p}} = \frac{PM}{RT} = \frac{10^5 \times 30}{8314 \times 1873} = 0.193 \text{ kg/m}^3$$

Substitution in Equation 9.5 gives:

$$f_{\mathrm{v}} = \frac{(1/1300)}{((1+12)/0.193)} = 1.14 \times 10^{-5}$$

We note that the calculated f_{v} is much less than 1, so our initial assumption is shown to be valid. This value is substituted into Equation 9.4:

$$\varepsilon = 1 - \exp\left(\frac{-3 \times 1.14 \times 10^{-5} \times 0.72}{2 \times 50 \times 10^{-6}}\right) = \mathbf{0.218}$$

If the combustion chamber is ten times larger, then the only change is the mean beam length, which also increases tenfold to 7.2 m. The densities and volume fraction depend only on the stoichiometry, not on actual flowrates, so the emissivity of the enlarged chamber is

$$\varepsilon = 1 - \exp\left(\frac{-3 \times 1.14 \times 10^{-5} \times 7.2}{2 \times 50 \times 10^{-6}}\right) = \mathbf{0.915}$$

Clearly there is a large increase in the emissivity of the combustion gases. This has great implications when scaling-up designs of combustion systems. In large pulverised coal combustors, such as those used for power generation, the combustion gases can be generally assumed to be black.

9.3.3 Small particles ($d > 0.5\lambda$)

Soot particles are much smaller than the wavelengths of radiation which are important in radiation heat transfer. Scattering of radiation by soot particles therefore falls in the regime of Rayleigh scattering. Theory predicts that the degree of Rayleigh scattering is strongly dependent on wavelength, varying as $1/\lambda^4$, whereas absorption varies as just $1/\lambda$. (These dependencies are slightly different in practice, because scattering and absorption are also dependent on the refractive index, which is itself a weak function of λ.) At the long wavelengths of interest in thermal radiation problems, scattering can be safely neglected, and attenuation is assumed to be due only to absorption.

The general outcome of the analysis is that the expression for the absorption coefficient K in Equation 9.2 is modified by a factor which includes dependence on the wavelength of the radiation and the complex refractive index (m):

$$K = \frac{3f_{\mathrm{v}}}{2d} \times \frac{24\pi d}{\lambda} f(m) = \frac{36 f_{\mathrm{v}} \pi}{\lambda} f(m)$$

We can define $A = 36\pi f(m)$ and simplify the algebra to give:

$$K = \frac{Af_v}{\lambda}$$

Note that K is now wavelength-dependent, so substitution of this expression into Equations 5.2 to 5.4 gives us an equation for the monochromatic emissivity and absorptivity:

$$\varepsilon_\lambda = \alpha_\lambda = 1 - \exp\left(-\frac{Af_v L_m}{\lambda}\right) \tag{9.6}$$

This must be intergrated using Equation 1.10 and Wien's Approximation (Equation 1.14) to give the total emissivity and absorptivity:

$$\varepsilon = \frac{\int_0^\infty \varepsilon_\lambda W_{b\lambda} \, d\lambda}{\int_0^\infty W_{b\lambda} \, d\lambda} = \frac{\int_0^\infty \left(1 - \exp\left(-\frac{Af_v L_m}{\lambda}\right)\right) \frac{2\pi hc^2}{\lambda^5} \exp\left(-\frac{hc}{\lambda kT}\right) d\lambda}{\int_0^\infty \frac{2\pi hc^2}{\lambda^5} \exp\left(-\frac{hc}{\lambda kT}\right) d\lambda}$$

$$= 1 - \frac{\int_0^\infty \frac{1}{\lambda^5} \exp\left(-\frac{Af_v L_m kT + hc}{\lambda kT}\right) d\lambda}{\int_0^\infty \frac{1}{\lambda^5} \exp\left(-\frac{hc}{\lambda kT}\right) d\lambda} \tag{9.7}$$

These two integrals are both in the form $\int_0^\infty \lambda^{-5} e^{-B/\lambda} d\lambda$, where B is a constant, and can be solved using the substitution $x = B/\lambda$ to give:

$$\int_0^\infty \lambda^{-5} e^{-B/\lambda} d\lambda = B^{-4} \int_0^\infty x^{-3} e^{-x} dx \tag{9.8}$$

Application of Equation 9.8 to Equation 9.7 gives:

$$\varepsilon = 1 - \frac{\left(\frac{kT}{Af_v L_m kT + hc}\right)^4 \int_0^\infty x^{-3} e^{-x} dx}{\left(\frac{kT}{hc}\right)^4 \int_0^\infty x^{-3} e^{-x} dx}$$

$$\varepsilon = 1 - \left(\frac{hc}{Af_v L_m kT + hc}\right)^4 = 1 - \left(1 + \frac{Af_v L_m kT}{hc}\right)^{-4} \tag{9.9}$$

The integration assumes that the weak dependence of A on wavelength (via the refractive index) is negligible. Equation 9.9 gives an algebraically straight-forward calculation of emissivity. The difficulty lies in providing values for f_v and A, both of which depend on the amount of soot being produced in the flame. The formation of soot depends in turn on the local and overall

Table 9.1 Experimental values of *A* for various combustion conditions

Fuel	A
Propane	4.9
Acetylene	4
Coal	3.7–7.5
Oil	6.3

stoichiometry and temperature. Soot concentrations are not uniform, so there is considerable spatial variation in both parameters. Values for f_v and A are best obtained from empirical data in the literature. Typical values for f_v in combustion gases are in the range 10^{-8} to 10^{-4}; for most industrial applications a value of f_v above about 10^{-6} will give gases which are essentially black. Experimental values of A for various combustion conditions are given in Table 9.1. A varies over a relatively small range, and a value of 5 can usually be taken as a good estimate if a literature value cannot be found. Any uncertainty in A is generally small compared with uncertainties in the value for f_v.

As indicated above, radiation from soot in flames is normally much greater than from the gases, such that gas radiation can be ignored. For small flames or flames with low soot concentrations, we can sum the contributions from the soot and the gases and then subtract a published spectral overlap correction of the type used for CO_2/H_2O mixtures.

Further Reading

Comprehensive texts:
Siegel, R. and Howell, J. R.: Thermal Radiation Heat Transfer (3rd ed.), Hemisphere Publishing, 1992.

Modest, M. M.: Radiative Heat Transfer, McGraw-Hill, 1993.

View Factors
Howell, J. R.: A Catalog of Radiation Configuration Factors, McGraw-Hill, 1982.

Applications
Rhine, J. M. and Tucker, R. J.: Modelling of Gas-Fired Furnaces and Boilers, British Gas/McGraw-Hill, 1991.

General heat transfer texts
McAdams, W. H.: Heat Transmission, McGraw-Hill, 1954.

Holman, J. P.: Heat Transfer (7th ed.) McGraw-Hill, 1992.

Kreith, F. and Bohn, M. S.: Principles of Heat Transfer (5th ed.) PWS Publishing, 1997.

Winterton, R. H. S.: Heat Transfer, OUP, 1997.

Index